INVESTIGATIONS INTO THE ORIGIN OF LANGUAGE AND CONSCIOUSNESS

BOSTON STUDIES IN THE PHILOSOPHY OF SCIENCE

EDITED BY ROBERT S. COHEN AND MARX W. WARTOFSKY

VOLUME 44

TRÂN DUC THAO

INVESTIGATIONS INTO THE ORIGIN OF LANGUAGE AND CONSCIOUSNESS

Translated by

Daniel J. Herman *and* Robert L. Armstrong

D. REIDEL PUBLISHING COMPANY

A MEMBER OF THE KLUWER ACADEMIC PUBLISHERS GROUP

DORDRECHT / BOSTON / LANCASTER

Library of Congress Cataloging in Publication Data

CIP

Trân, Duc Thao.
Investigations into the origin of language and consciousness.

(Boston studies in the philosophy of science; v. 44)
Translation of: Recherches sur l'origine du langage et de la conscience.
Includes index.
1. Language and languages–Origin. 2. Psycholinguistics.
I. Title. II. Series.
Q174.B67 vol. 44 [P116] 501s [401'.9] 83-17726

ISBN-13: 978-94-009-6238-5 e-ISBN-13: 978-94-009-6236-1
DOI: 10.1007/978-94-009-6236-1

Published by D. Reidel Publishing Company,
P.O. Box 17, 3300 AA Dordrecht, Holland.

Sold and distributed in the U.S.A. and Canada
by Kluwer Academic Publishers,
160 Old Derby Street, Hingham, MA 02043, U.S.A.

In all other countries, sold and distributed
by Kluwer Academic Publishers Group,
P.O. Box 322, 3300 AH Dordrecht, Holland.

This edition of Trân Duc Thao's *Recherches sur l'origine du langage et de la conscience* (Paris: Editions sociales, 1973) has been edited by Carolyn R. Fawcett and Robert S. Cohen.

TABLE OF CONTENTS

EDITORIAL PREFACE

Trân Duc Thao, a wise and learned scientist and an eminent Marxist philosopher, begins this treatise on the origins of language and consciousness with a question: "One of the principal difficulties of the problem of the origin of consciousness is the exact determination of its beginnings. Precisely where must one draw the line between the sensori-motor psychism of animals and the conscious psychism that we see developing in man?" And then he cites Karl Marx's famous passage about 'the bee and the architect' from *Capital*:

> . . . what distinguishes the worst architect from the best of bees is this, that the architect raises his structure in the imagination before he erects it in reality. At the end of every labor process, we get a result that already existed in the imagination of the laborer at its commencement. (*Capital*, Vol. I, p. 178, tr. Moore and Aveling)

Thao follows this immediately with a second question: "But is this *the most elementary* form of consciousness?" Thus the conundrum concerning the origins of consciousness is posed as a circle: if human consciousness presupposes representation (of the external reality, of mental awareness, of actions, of what it may), and if this consciousness emerges first with the activity of production using tools, and if the production of tools itself presupposes representation - that is, with an image of what is to be produced in the mind of the producer - then the conditions for the origins of human consciousness already presuppose the very form of consciousness which they are supposed to explain.

It was reasoning of this sort that led Hegel to develop a theory of the mind which made consciousness itself the presupposition, the very originating condition, for production, indeed for the existence and the coming-into-being of the objects of consciousness. These objects, or perhaps better said, these *objectifications*, become the means by which consciousness then becomes aware of itself, as self-consciousness. But if this Hegelian primacy of consciousness is plainly an idealist solution to the problem, what would the alternative materialist solution be? Indeed, if we take Marx at his word here, then he would appear to propose idealist presuppositions for the explanation of the elementary forms of consciousness, and this would then seem to be in stark contradiction to the materialist, *praxis*-oriented account which Marx and Engels have developed in their historical materialism.

There is, however, another way to ask the question, and it is in fact compatible with the historical materialist account: *deny* that the architect, in contrast to the bee, represents the *most* elementary form of consciousness, and then propose that representation, as the essential precondition of human consciousness, itself has its genesis, which is to be found in still more elementary forms of *pre-representational* consciousness, which exists prior to the fully human forms of production, and prior to the making and use of tools. This is the path taken by Trân Duc Thao in this incisive and imaginative study. He turns, as we see, to the pre-hominid development of language, to that crucial (and dramatic) moment which is, in his account, the act of signifying an object, i.e., the development of the primitive linguistic sign. He begins, therefore, in the First Investigation, with his analysis of the indicative gesture.

With that, we are launched into one of the most sophisticated and anthropologically informed treatments of the origins of language and of consciousness, set forth within a broadly Marxist framework. Trân Duc Thao proceeds from the indicative gesture, which is of course at once the elementary linguistic sign, to the development of self-recognition and to the possibilities of self-reference on the basis of reciprocal recognition of the other, and then to the forms of 'echoic representation' - his interpretation of the classical dialectic of self-consciousness within the terms of pre-hominid and early hominid *praxis* and social interaction. He pursues the stages of this dialectic from what he neatly characterizes as 'sporadic cognizance' to collective cognizance, and thence to individual cognizance, in a way reminiscent of both Vygotsky's and Wallon's theories of cognitive development in the child; and then he traces the praxical origins of the 'ideality of consciousness'.

In the Second Investigation, Trân Duc Thao develops his theory of signification, the crucial relation of meaning to instrumental activity, and explores the progress from the indicative sign to genuine representation. "The *qualitative leap*" he writes "is realized only with the transcendence of the present perception through the beginning of representation" (p. 59). In a detailed analysis, combined with a close discussion of the work of Piaget and that of Gvosdev, he traces the beginnings of representation in the child; and we see thereby the development of the representational sign and the functional sentence, in both a phylogenetic and an ontogenetic context. Throughout this impressive analysis of the development of language and meaning, Trân Duc Thao embeds it within an account of the forms of *praxis* in which such a linguistic development might plausibly have taken place: in the transition from the production of the 'instrument' to the production of the 'tool' as such, from *Homo habilis* to *Homo faber*.

With his Third Investigation, a striking essay on 'Marxism and Psychoanalysis on the Origins of the Oedipal Crisis', Thao again proceeds with a socio-historical reconstruction of the *genesis* of the Oedipal. Here, as we see, he questions Freud's Oedipal theory, not indeed in the post-structuralist way of Deleuze and Guattari in their *Anti-Oedipe*, but rather in terms of a historical materialist critique and reconstruction of the hypothetical origins of Oedipality in a pre-Oedipal stage. He traces this to the social, rather than biological, development of the transition from what may be called animal 'jealousy' to the suppression of this 'zoological individualism' as a condition for the formation of the first social group necessary for the beginning of human production; and the further drama of human history follows, the 'reawakening of jealousy' and the emergence of the Oedipus Complex as a later stage, developing along with the transition from the communalization of women to the pairing family.

In all of these 'Investigations', Trân Duc Thao weaves a rich and complex argument from the strands of anthropology, linguistics, archeology, cognitive and developmental psychology and epistemology. He has, of course, his profound familiarity with the thought of Marx and Engels, Hegel, and an impressive list of contemporary thinkers, European, American and Soviet. Thao moves, often magisterially, sometimes daringly and riskily, among these fields, with a striking suppleness of mind and with great originality. He has written a work which should stimulate a new and deeper approach to the origins of human consciousness, the origins of all of us.

*

Trân Duc Thao is author of the seminal series of studies of Husserl and Marx which were published in the 40s and 50s, and which have been gathered into his modern classic, the *Phénoménologie et matérialisme dialectique* (Editions Minh-tân, Paris 1951; soon to appear in a fine English translation by Daniel J. Herman and Donald V. Morano within our *Boston Studies*). The present work, written in the 60s and early 70s, and published in Paris in 1973 by Editions sociales, continues and modifies the account of the origin of consciousness given in the essay on dialectical materialism in the earlier book. Whether a phenomenological mode of analysis is thoroughly replaced in the book before us, or perhaps is *aufgehoben* by a socially articulated, which is to say by a historical materialist, analysis, or whether there are still phenomenological and also biological elements, will be debated by Trân Duc Thao's attentive readers. To say with Marx, once again, that "language is practical consciousness" is also to welcome Thao's instructive exploration of a social and evolutionary theory of what is latent as well as manifest in human nature.

*

We are most grateful for the exemplary achievement of Professor Daniel J. Herman and Dr. Robert L. Armstrong in their translation of the *Recherches sur l'origine du langage et de la conscience.*

We are also pleased that circumstances have enabled us at last to present the pioneering work of this distinguished philosopher from Viet Nam to the English-reading scientific public in our own country and world-wide.

Center for Philosophy and History of Science, ROBERT S. COHEN
Boston University

Department of Philosophy, MARX W. WARTOFSKY
Baruch College of the City University of New York

The history of philosophy, *ERGO:*

kurz,* the history of cognition in general

the whole field of knowledge

Greek Philosophy indicated all these moments

The history of the separate sciences
" " " the mental development of the child
" " " the mental development of animals
" language N.B.:
+ psychology
+ physiology of the sense organs

these are the fields of knowledge from which the theory of knowledge and dialectics should be built

* briefly

From: V. I. Lenin, *Collected Works*, 47 vols. (London: Lawrence and Wishart; Moscow: Foreign Languages Publishing House, 1960–1980). vol. 38: *Philosophical Notebooks*, pp. 352–353.

FIRST INVESTIGATION

THE INDICATIVE GESTURE AS THE ORIGINAL FORM OF CONSCIOUSNESS

One of the principal difficulties of the problem of the origin of consciousness is the exact determination of its beginnings. Precisely where must one draw the line between the sensori-motor psychism of animals and the conscious psychism that we see developing in man?

It would seem natural to date the beginning of consciousness with the very beginning of humanity, with the appearance of the first tools whose production already implies a previous representation of their form in the head of the subject who produced them. Marx says,

> . . . What distinguishes the worst architect from the best of bees is that the architect builds the cell in his mind before he constructs it in wax. At the end of every labour process, a result emerges which had already been conceived by the worker at the beginning, hence already existed ideally.[1]

Thus the first men who appeared in the pre-Chellean era already had an elementary form of consciousness as part of the activity of production. But is this *the most elementary* form of consciousness? If we allow that it is, we must explain it directly in terms of the development of tool-using activity as it begins to appear among the apes. Now, this seems hardly possible, for the use of natural objects as implements is only a *labor of adaptation* to the present situation. The form of the natural instrument can change each time so long as it somehow satisfies the immediate need. The representation of the tool, on the contrary, implies the image of a stable form, fixed in itself once and for all. The habit of tool-using activity in natural conditions will obviously make the animal-subject more skillful, but we cannot see at all how, by itself, this habit could bring the animal to transcend the immediate exigencies of its present situation and raise itself to the ideal representation of a typical form that would allow it to pass to tool production.

It is evident that the distance between the work of adaptation, as we find it sporadically in the anthropoids, and the ideal image of the tool which Pithecanthropus must have already had in his head in order to enable him to undertake the labor of production, is far too great to be immediately overcome. We must, therefore, look for the beginning of consciousness at an intermediary stage, before the appearance of the most ancient man.

Marxist anthropological research has established the existence of a *prehominid* stage, notably represented by the remains of Australopithecus. "The Australopitheci were highly developed primates who walked with two feet on the ground and in the evolutionary process of adaptation to difficult environmental conditions, had succeeded in systematically using various natural objects as instruments."[2] This is a stage where the ape had already risen above animality, strictly speaking, by assuming the habit of the work of adaptation, without having yet attained the form of production characteristic of human society. It is thus at this level that we must look for the original form of consciousness, such as it arose in the course of the development of tool-using activity, which, rooted in animal evolution, brings about the transition to humanity.

Consciousness must first of all be studied in its 'immediate reality': language understood naturally in its general sense as gestural and verbal language. Language is originally constituted on the very activity of adaptive labor, starting at the anthropoid level. "First comes labour," says Engels, "after it, and then side by side with it, articulate speech – these were the two most essential stimuli under the influence of which the brain of the ape gradually changed into that of man . . ."[3]

The apes we know, strictly speaking, have no language. Their various means of expression – gestures, cries, etc. – refer primarily to the emotional, essentially biological aspect of their situation. These means can also serve as signals for determinate behaviors, but, taken strictly, they do not have any meaning signifying an object, which proves that in the ape there is no conscious relation to the object such that it is expressed precisely in language. Even the anthropoids lack the most elementary linguistic sign: the indicative gesture. This can be seen in the following story told by the owner of a gorilla:

> A piece of filleted beef-steak had just come from the butcher. Inasmuch as occasionally I gave him a small mouthful of raw beef, a small piece of the coarser part of the steak was cut off, and I gave it to him. He tasted it, then gravely handed it back to me. Then he took my hand and put it on the finer part of the meat. From that I cut off a tiny piece, gave it to him, and he ate it.[4]

We are obviously dealing here with a very intelligent animal. But this account shows precisely that animal intelligence does not reach the level of consciousness. If the gorilla had had *conscious* perception of the steak, he would have simply *indicated* the part of the meat that he wanted by stretching out not his finger, to be sure, but at least his hand. The indicative gesture marks the most elementary relation of consciousness to the object as external object. Naturally the animal perceives the external object, but for him the externality of the object is not detached from his own sensori-motor organization. In other words, the image provided by his perception of the object can be defined only in terms of the potential action of his own body – extended, more or less, by other bodies that he can manipulate. For him, then, the object is what he can actually *touch*, directly or through an intermediary: this is exactly what the gorilla does with the piece of meat through the intermediary of his mistress's hand, which somehow extends the movement of his own hand.

And it is precisely because the psychic image of the object presents itself contiguously with the potential movements already more or less set in operation in the body of the animal that he is incapable of *indicating* the distant object, even though the distance of the object is minute. In other words, he does not have the concept of distance as such, so that his perception, though it is of the external object does not include awareness of the externality of the object. Yet, the most essential characteristic of human perception as *conscious* perception is to recognize precisely the external object *as external*, which implies the awareness of distance *per se*, the concept of the relation of the externality of the object to the subject.

This is obviously the basis of the '*naïve realism*' of any mentally healthy person of which Lenin speaks, namely the conviction "that things, the environment, the world exist *independently* of our sensation, of our consciousness, of our *self* and of man in general."[5] And the meaning of the indicative gesture is precisely to express the relation of objective externality, in which consists the fundamental intentionality of consciousness as consciousness *of the object* as opposed to the simply sensori-motor psychism of the animal.

As we see it in man, the indicative gesture appears in two forms. The child who shows his mother a jar of jam can either point his index finger directly at the object, or he can make a circular movement, the hand being raised first toward the mother and then toward the object. In the adult, the gesture is made almost exclusively in the form of a straight line, but the circular-arc form can also appear in some emotional situations. Thus, if I show the door to someone, my finger will naturally point to the door. But, if I am really in a rage, I could begin possibly by lifting my finger toward the person in question in order to direct him toward the door. Since we know that the primitive forms of behavior reappear in moments of emotion, we can think that it was the circular-arc motion that first appeared in phylogenesis. On the other hand, that form continues to play its role together with the straight line gesture. When I point my index finger directly at an object, I normally look at the person to whom I am making this gesture. If we consider the movement in its finished form, the eyes, while scanning space, must come back to the indicated object. This is precisely what is called 'pointing by looking'.

Thus we can consider the circular arc form as the fundamental form. It presents a striking analogy to the movement of the gorilla cited earlier: "Taking my hand, he put it on the best part of the meat." We are dealing here with an act of *guidance* which differs from the circular indicative gesture only in that the human subject keeps himself at a distance, while contiguity is necessary for the anthropoid: the gorilla takes the hand and guides it toward the desired piece of meat. The original form of the indicative movement thus can be defined as *guidance at a distance*. As such it is obviously derived from the *contiguous guidance* used by the anthropoids. How could the transition from one to the other have occurred? It is fairly clear that the development of adaptive labor played a determining role here. During the recession of the tropical forest in favor of the steppe, toward the end of the Tertiary Period, when many species of apes disappeared, only those who could adapt themselves one way or another to the difficult conditions of the new environment were able to survive. Some of them, quite gifted from the psychic point of view, developed the use of natural instruments, notably stones and branches serving as sticks. As apes normally live in a group, labor was collective, which required a minimum of coordination concentrated on any one object of labor. The workers could not guide themselves by the hand since their hands were already occupied in holding an instrument. Moreover, since the usefulness of the instrument consisted in enlarging the field of action of their own bodies, each one had to keep a certain distance from the others in order to avoid interfering with them. Contiguity thus was broken and the guidance gesture was necessarily performed at a distance.

The principal resource of the steppe consisted in the animals of the Ungulata order. From the archeological excavations at the camps of the Australopitheci we know that the prehominids hunted big game like the antelope and the giraffe. They preferred to attack young or very old animals, that is to say, the weakest.[6] We can think that their ancestors, beginning with the anthropoids, already had conducted collective hunts of the same kind. At first, coordination was rather shaky. The guidance gestures were first performed at a short distance; in other words, they first concerned only the adjacent hunters. Then the motion of the hand extended progressively, and ended up by encompassing the whole group of hunters who concentrated their efforts on the weakest animal of the ungulate herd. In this way, the broad circular-arc form of gesture originated, which we have recognized as the primitive form of the indicative gesture.

Such an acquisition constitutes qualitative progress that was probably realized in the transition from the anthropoid to the prehominid. In fact, in the prehominids, so far as we can judge from the remains of Australopithecus: "instrumental activities, which previously exhibited a contingent character up to a certain point, have acquired the character of biological necessity conforming to a law."[7] It is obvious that adaptive work could not take the form of a signaling behavior that makes possible a concentration of collective efforts on the same object. This is precisely what the indicative gesture in a circular arc does, as the result of the evolution of the guidance movement which began with the anthropoid stage.

The whole process has been accomplished so far within the sensori-motor framework of animal psychism; nowhere have we seen the intervention of consciousness, which in fact has yet to appear. It is only starting with the already acquired *objective form* of indication that the *subjective form* is constituted, which defines the first *intentional* relation of subject to object, as the *original consciousness* of the object.

Indeed, once the structure of the gesture is established, the subject applies it to himself. In other words, *he points out the object to himself*. It is this movement that we witness in children when they are looking at a particularly interesting scene. Once I observed a little girl of 18 months sitting alone by the window and looking into the street. At a certain moment she lifted her arm and pointed her index finger toward the street. Her gesture was obviously addressed only to herself, for I was sitting at the other end of the room, and she had had her back turned toward me for quite a while; she was pointing out the scene to herself.

The indicative gesture to oneself naturally derives from the gesture we use in pointing things out to others. However, it contains a difference which confronts us with a very fundamental problem. The indicative gesture to others, which we have defined in its original form as a guidance act at a distance, implies, in fact, at least two subjects, one guiding and the other guided, separated by a certain distance. In the case of the indicative gesture to oneself, we have only one subject, both guiding and guided, so how under these conditions can guidance be accomplished '*at a distance*'? Evidently, the movement is possible only if the subject *considers the distance*, so to speak, *in relation to himself*. Moreover, it is what we more or less feel within ourselves if we make this gesture or some other signifying gesture to ourselves. The phenomenon is quite evident in the case of the 'internal dialogue' when

I address myself in the second person: I obviously place myself in the position *of another, who is precisely myself*, and it is from that point of view that I address myself *to myself as another*. But how is this possible?

We must return to the original indicative gesture in order to examine a relation which has been ignored so far for the sake of clarity of exposition: this is the relation of reciprocity. In the activity of collective labor, the workers point out to each other the object of their common efforts. Each is thus, alternatively, or even simultaneously, the giver and the receiver of the indication, both the one who guides and is guided. The workers see each other reciprocally in this double function. In other words, each sees in the other a being similar to himself, making the same gesture or rather he sees the other as *another self*. And it is precisely because he sees himself in the others that the enduring image of the social environment allows him, when alone, to take the point of view of these others who are his other self in order to guide himself 'at a distance' toward the object, in other words, to point out the object to himself.

" ... a man," says Marx, "first sees and recognizes himself in another man. Peter only relates to himself as a man through his relation to another man, Paul, in whom he recognizes his likeness."[8] This is true from the very beginning to the time when prehominid man, having arrived at the upper limit of animality through the habit of adaptive work, made the transition from sensori-motor psychism to the original form of human consciousness.

In the movement of reciprocal indicative gestures, prehominid workers reflect themselves, so to speak, in sending one another the same gesture, and see themselves in the others 'as in a mirror'. At the stage at which we have arrived, of course, perception consists only of sensori-motor images which are not conscious. From this stage on, however, the subject recognizes himself in his own image. This can be seen in Köhler's narration of his experiments on the behavior of chimpanzees with a mirror.

> When we gave the chimpanzees a hand-mirror for the first time, they looked into it and at once became intensely interested. Each one wanted to look ... [their] interest ... did not decrease but remained and became one of the most popular and permanent of their 'fashions'. ... they mirrored themselves in anything at all available for the purpose: ... above all, in pools of rain water. I have often observed Tschego for long [periods] at a time sunk in contemplation of her own reflection in a pool. She played with it: bent far over it and drew back slowly, shook her head backwards and forwards, and made all kinds of grimaces, over and over again.[9]

Thus the anthropoid recognizes himself in his own image, in the sense, of course, of a purely sensori-motor recognition. In other words, he perceives the relation of resemblance between his own body and its image in the mirror. Thus when prehominid workers send one another the same indicative gesture, each one sees his own gesture in the symmetrical gesture of the other, or rather he "sees himself in the other as in a mirror."

But where can this practice of returning gestures or reciprocal reflection lead? Here we must specify further the function of the indicative gesture and complete its description in order to draw out the consequences of reciprocity.

When defining the gesture as a guidance at a distance we have so far insisted on its form. In reality the guidance movement does not consist in simply tracing a direction, it has essentially the function of a *call*. The indicative gesture, as distance guidance, is a *call for work on the indicated object*. Now, as a call, it is naturally completed by the normal form of a call, the vocal form. Thus when children, at the beginning of their second year, are starting to make the indicative gesture, they accompany it with the exclamation 'Ah!'. We know that the sound '*ak*' is emitted by apes when they see an unusual object.[10] Obviously this is only an animal cry which relates to the disquieting character of the situation and not to the object itself in its own reality. But at the prehominid stage, when the cry accompanies the indicative gesture, it takes, thereby, the meaning of object. It becomes the *exclamation* that defines the original form of verbal language and indicates the object as an object of work: the 'this here!'. The indicative gesture thus contains two moments, the gestural moment and the exclamatory moment.

We must notice, though, that the form of reciprocity is even more marked in the exclamatory component than in the gestural one. We do, in fact, find this reciprocity of cries as early as the animal stage. Mammals living in a group take up each other's cry as *in an echo*. In a group of chimpanzees, it is enough for one animal, seeing or believing himself attacked, to let out a cry of indignation for the others to repeat it immediately and rush to a collective attack.[11]

When, therefore, at the prehominid stage, the cry becomes an exclamation, these exclamations answer one another like an echo, and considerably reinforce the image of the prehominid that each one sees in the others. The workers call themselves to the work-object by means of gesture and voice, and each one sees himself in the other as in a mirror and hears himself in the others *as in an echo*.

Until now, the subjects have merely returned the sign to one another, without any of them addressing it to himself. Therefore we have not yet reached the level of consciousness. However, once the structure of reciprocity is acquired, situations occur in which the gesture is necessarily deflected in a new direction. If, for example, in a hunt one of the hunters lags behind the others, and the others call him by indicating the game with a gesture of the hand in a circular arc, it is evident that he does not have to call them in the same manner, since he is indeed lagging behind; rather it is up to him to catch up and rejoin the group around the object indicated. It is true that once the structure of reciprocity has already been established in the sensori-motor setting, the subject has a tendency to return automatically to the others the sign he has just received. But since the situation is different, the gesture can no longer be made in the simple form of symmetry. At the very moment when the subject begins to return the gesture to the others, he finds himself in contradiction with his own position as a lagger. The result of this is that the barely outlined gesture is immediately absorbed in the sign sent out by the group, so that, in fact, it is *for himself* that the subject repeats this call, which amounts to saying that *he calls himself* to rejoin the others.

This calling to himself which takes up the call of the others is immediately realized on the vocal level, for the exclamations which answer one another blend like a choir. When we sign in unison, our own voice seems to come as much from the others as from ourselves, and we hear their voices equally resounding in us. The case is the same for the prehominid subject who takes up the exclamations of his group; everything happens as if his voice came to him at once from others and from himself. Of course, in the general case, confusion is limited by the fact that individuals effectively point out the object to one another so that the exclamations remain distinct to a certain extent. But in the situation of the lagger, the moment of the gesture to the others tends to disappear and the subject hears his own voice *coming from the others as in himself*. As a result, the exclamation which he emits is *identified* with those the others address to him, so that his own call *comes back to himself* and is, in fact, addressed to himself.

The process thus realized on the vocal level is extended to the gestural level. It is true that we are faced here with a difficulty, for the indicative gesture in the circular-arc form can only be addressed to others, and the identification with others' gestures thus implies the creation of a new form. The subject begins by answering in the original form which consists in raising the hand in the direction of others in order to carry it back to the object. But as a lagger he does not, in fact, have to indicate the object to others, and since he has already identified himself with them by voice, the result is that his barely outlined gesture is immediately confused with theirs and is absorbed in the part common to two symmetrical gestures, namely the phase where the hand motion of the others goes back from himself to the object.

In other words the gesture of the subject *returns upon himself* in order to go directly *from himself toward the object*. We see that the form of the straight-line gesture is constituted as the synthesis of the two reciprocal moments of the circular-arc form, the first when the subject is the giver, and the second when he is the receiver of the gesture. Thus he is now in one and the same gesture *both the giver and the receiver*. In short, by virtue of his situation as a lagger, the indicative gesture as a whole, gesture and voice, returns to himself, in such a way that the subject, starting with the others *with whom he identifies, points out the object to himself.*

Now it is this very dialectic, where the reciprocity of the sign is absorbed in the form of identity, which engenders the structure of *lived experience* wherein the relation to oneself constitutes *consciousness*. Consciousness appears identically as *consciousness of the object* and *consciousness of self*. As consciousness of the object, it is the *image of the object* posited as external to itself. As consciousness of self, it is the *image of that image*, or *the image of itself in itself*. Of course, when we speak of consciousness as an image of the object, we mean it in the active sense, as a productive act of image. Now, in the indicative gesture to oneself, such as it has just arisen, the subject gives himself an image of the object in its not only most elementary but also most fundamental determination, that is, its objective externality, a determination where the object appears as the 'this here!'. And at the same time, this act, which constitutes the perceptive image of the object, has its own image in the gesture of the others with whom he *identifies*, so that this image of himself, which the subject finds in the others, presents itself as *within himself*. Here we have an *image of the object*: this image is *accompanied by the image of itself in itself* which consists of the *consciousness of the object*, this consciousness being *experienced in itself*. The relation to the self arises as a *result* of the relation with the other, exactly as Marx says: "Peter only relates to himself as a man through his relation to another man, Paul, in whom he recognizes his likeness."

Consciousness, which thus arose in a rather particular situation, still presents itself only in a *sporadic* manner as a *flash of consciousness*. This flash was sufficient, however, to fix the form of the straight-line gesture, which was to become the predominant form of indicative gesture to others. In fact, the original circular-arc form is of rather limited use since it actually implies an urgent work situation; faced with a herd of antelopes, for example, prehominid hunters had collectively to attack the weakest animal, and it is the motion of the hand in a circular-arc which carries the whole group in the same direction. We can still find this sign today in the classical gesture of a unit commander about to mount an attack: the hand makes a large gesture in a circular arc which envelopes the men and which sweeps them in the direction of the enemy. We can see a plastic representation of it in Delacroix's painting *La liberté conduisant le peuple* [*Liberty Leading the People*, 1831. Paris, Louvre.]: her right hand which carries the flag, is on the verge of making a large semi-circular gesture, while her eyes, turned toward the fighting men, accentuate her pointed gaze.

It is evident that such a sign would have been much too strong in everyday life, when it was just a matter of the prehominids drawing each others' attention to a more or less interesting object and not preparing themselves for an attack. The indicative gesture in this case then must be performed in an *attenuated* form.

Now it is precisely such a form that we find in the straight-line gesture. While the indicative sign of the circular-arc form brings the receiver himself to the indicated object, the straight-line sign simply directs his glance, which follows the extension of the hand gesture stretched toward the object. The person to whom the sign is addressed concerns himself with it only if there is a sufficiently powerful motive. We can see that the straight-line form allows for a considerable extension of the field of application of the sign which henceforward can serve to indicate not only the current work-object, but also anything interesting that may be a possible work-object.

Obviously, at the beginning, this extension of the domain of indication concerns only particularly interesting objects, capable of strongly exciting the interest of the group. This schema, however, is progressively strengthened in itself by virtue of the conscious moment that it implies. Since the gesture now goes directly from the subject to the object and is consequently found to be meaningful for others as well as for himself, it is clear that in the dialectic of reciprocity each subject, by returning the sign to others, addresses it at the same time to himself. In other words, it is equally for himself that he takes up again the sign sent by others, so that he addresses it to himself *starting with others* and sees himself at the same time in them as in *other selves*.

"A man," says Marx, "first sees and recognizes himself in another man." [12] The *cognizance* [*prise de conscience*: grasp of consciousness] of the indicative gesture began *sporadically* in the lagging prehominid hunter [13] who repeated to himself the call of the others and *recognized himself in them*. Cognizance now develops into a *collective cognizance* where all the individuals of the group *recognize each other in the others*, since they at the same time address each to himself the sign which they return to one another, so that all merge in one and the same gesture where each sees himself in the others as in himself. The sign, consequently, which thus appears to each individual as *experienced in himself* insofar as he is part of the action of the group and is identified with it, is somehow sustained by the social relation itself. In other words, *the sign has been internalized by the group*, in such a way that it becomes for the group *available experience* that subsequently can be used at will; in other words, it can be applied not only to particularly interesting objects, but also to any more or less interesting object *in general*. Cognizance will be complete when the act of the group can be reduced to its own *enduring image* in the individual, so that the sign is internalized for the individual himself, insofar as he has gathered in himself the identified form of social reciprocity.

We can already detect in the anthropoids the existence of an enduring image of the group whose influence is evident in cases of the isolation of the individual. Thus when a chimpanzee is separated from his companions, he begins to yell, cry and angrily hurl himself against the walls of his enclosure. The first few days he even refuses food altogether.[14] The *social call*, which subsists in a permanent form in the individual psyche, has evidently only an emotional value in this case. But the moment that the linguistic sign signifying its object appears, the enduring image of the group necessarily invokes this objective content. It is a fact of common experience that we constantly feel around us the presence of our familiar social environment, and this image essentially implies the typical form of gestures and words of people we know. The meaning of the world in which we live is defined precisely *for us* by what this social image *tells* us. And we perceive and recognize present persons through this 'model' which envelopes the whole complexity of human relations and all the richness of lived experience.

The stage at which we have arrived, the dialectical moment when animal intelligence raises itself to the original form of consciousness, the enduring image of the others, in its significant form, as yet only implies the traits which have been fixed in reciprocal indications by both gesture and voice. The prehominid keeps within himself this stylized image of a multitude of indicative gestures which always seem to call him in chorus to work on the object. And even though the group actually emits this sign only in determinate circumstances, each individual nonetheless constantly perceives the others through this 'model.' By virtue of the already acquired reciprocal structure the subject responds in the very form just described; in other words, he returns the sign by the straight-line gesture to the enduring image of the others and at the same time addresses it to himself. We see that the gesture here is sustained by the already internalized social relation so that it somehow arises from the very play of reciprocal images where the lived experience of the subject himself is constituted. In other words, *the sign has been internalized not only for the group but for the individual as well.* It thereby becomes for the individual *available experience* which tends to be applied *to every perceived object in general*, since the constant presence of the image of the group, under the form of a constantly repeated chorus of indications, awakens the same gesture in the individual as soon as an object enters within the field of his sensori-motor perception. In *individual cognizance* mediated by the stylized image of the group, the *generalization of the sign* is completed. It stands out from the contingent circumstances of its appearance and is reduced to its general structure, such that it is disengaged from the activity of social practice, where the fundamental determinate nature of the object is reflected in its concrete universality as external reality independent of the subject – every object is a '*this here*'.

We see then that in its original activity, cognizance is realized in three moments. First of all a *sporadic* cognizance arises in a situation of lagging behind, as a flash of consciousness. Its function is to incite the subject to overcome his own slowness in order to place himself at the level of social action. Then comes the *collective* cognizance which permits a first generalization of the sign by making it available to the group outside of urgent work situations. Finally comes *individual* cognizance which completes the generalization of the sign by making it constantly available to the individual. From now on the subject can systematically utilize this sign, of which he has become conscious in himself, in order to act on himself and others, to mobilize and direct the energies of the group for the appropriation of the object. From the very beginning this is obviously the foundation of the *practical role of consciousness* which will greatly expand throughout the history of the *hominid* family.

Consciousness, as it has just arisen, already implies the individual form of the *self* [*soi*]. If we consider its content, however, it still remains simply collective. Thus, it does not in any way contain the form of the 'I' [*moi*]. The 'I' assumes a great deal of mediation which will be progressively constituted in the dialectic of social development. In the child, the 'I' does not appear until the second half of the third year, while the indicative gesture develops as early as the fourteenth month. At the birth of the prehominid, the gesture of reciprocal indication in the collective work of adaptation implies a complete assimilation among subjects who indicate to one another the object of their common efforts: consciousness thus appears as "mere herd-consciousness" or "sheep-like consciousness."[15] Lived experience is still 'anonymous'. It nonetheless implies the fundamental meaning of the *self*, as dialectical identity of the subject with his internal image which he constantly possesses in the enduring image of his group with which he identifies himself. And it is in this relation of self to self, where the self is experienced *in itself*, that the original relation of consciousness *to the object* is constituted: the indication to oneself experienced as an *intentional focussing* on the singular object in its objective externality, the '*this here*' of '*sense certainty*'.

The indicative gesture began in a material form: the gesture of guidance at a distance as it was formed in the development of adaptive work starting with the anthropoids. It is now completed in an ideal form as the intentional object within lived experience. Judging from appearances one could believe in some sort of 'transsubstantiation'. Idealism defines the ideal form of consciousness in its very substance as 'immaterial substance'. But this is, of course, to give in to appearances. In reality, the *material component always remains present*, in one way or another, in the ideal form. And its presence is a proof that the authentic *subject* of consciousness, the being who moves in the ideal motion of lived experience, is the real social subject in flesh and bone. "Consciousness [*das Bewusstsein*]," says Marx, "can never be anything else than conscious being (*das bewusste Sein*), and the being of men is their actual life-process."[16] Now the 'actual life-process' of men can be none other than their *material* behavior: "One could not separate thought from thinking matter. It is the subject of all changes."[17]

When I look at an object, I naturally don't have to point it out to myself in order to obtain the 'sense certainty' of its objective reality. But the material act of indication is revealed in the motion of the glance. The human eyes have an expression that we do not find in the animal. The animal *orients* his eyes toward the object, the human gaze *indicates* the object, to himself as well as to others.

As for the gesture of the hand, one rarely finds it externalized. It nevertheless remains present in the form of an internal outline. The outlined gesture, weak though it may be, is clearly as real and material as the finished gesture. We call it 'internal' simply because it is not seen from the outside, but its presence is demonstrated precisely in those cases in which it happens to be externalized. We have already noticed this gesture in the child, and the same is the case with the adult, though it occurs more rarely. On a boat once, just as we came into view of the coast, I saw a sailor alone at the bow lifting his arm and pointing his index finger toward land. He remained that way for a long moment, immobile as a statue. His gesture could only be addressed to himself: he pointed out the land to himself. We had just come out of a storm which had made the boat deviate from its course. It was a sailboat and there were only the three of us on board. The abnormal duration of the journey had undoubtedly made the sailor impatient to see land again. When the coast appeared, his emotion led him to finish the gesture which normally is barely outlined. By thus externalizing the gesture, he reinforced his *sense certainty* of the object, since it consists precisely in the experienced act of pointing out the object to oneself. And it is this very same thing which prolonged the externalization of the gesture for a moment.

The movement of the glance and the gesture of the hand, outlined or completed, are associated with the exclamation addressed internally to oneself: we feel it outlined, so to speak, in the movements of the throat and tongue, and, moreover, we also happen to externalize it as "Ah!" or "That!". Thus the material component, even though generally difficult to grasp by virtue of its outlined form, is, nevertheless an integral part of the experienced movement. Its necessary presence in the act of consciousness enables us to define it as *language*, or unity of the signifying act, as material sign, with what it signifies.

"Language," says Marx, "*is* practical, real consciousness."[18] Language is thus not simply the *expression* of thought or of consciousness. Language – which must obviously be understood both as gestural and verbal language – *is consciousness itself in its 'immediate reality'*. More precisely, consciousness is *the language that the subject addresses to himself*, generally in the outlined form of 'internal speech'. And as we have already noted, the outlined movement or internal movement is just as real, as material, as the externalized movement. This can be demonstrated by monitoring the bioelectirc currents at the muscular level. Moreover, when we think, we do indeed feel the action of the vocal organs and the hand. The subject is *conscious* of what he thinks, thanks to the internal perception, starting from the kinesthesia, and visual and auditory associations of the outlined motion of his gestures and of his voice. When we say that language expresses thought or consciousness, this simply means that formulated language expresses generally explicitly and in an externalized way the meaning expressed in an outlined and abbreviated manner in internal language.

But where could the meaning of internal language itself come from? It is evident that since consciousness is secondary to matter, its meaning as the meaning of internal language cannot, – at least essentially – derive from itself, but rather must come from objective reality. And as all meaning implies a language, we must indeed admit the existence of a language belonging to reality itself, prior to consciousness, from which consciousness draws its meaning. In other words, meaning must be constituted first of all in an objective form in the original movement of language as it sprang directly from the material relations of social life, what Marx called "*the language of real life*." And it is only on this basis that consciousness is established as "the direct efflux of . . . material behavior." "The production of ideas, of conceptions, of consciousness," says Marx, "is at first directly interwoven with the material activity and the material intercourse of men – *the language of real life*. Conceiving, thinking, the mental intercourse of men at this stage still appears as *the direct efflux of their material behavior*."[19] "The language of real life" as we find it here, is clearly the immediate expression of the movement of "material activity" and of the "material intercourse of men." These three moments: material activity, material intercourse and language of real life, constitute the "material behavior" of men whose consciousness or thinking is originally "the direct efflux . . . [*direkter Ausfluss ihres materiellen Verhaltens*]." *The language of real life is thus prior to consciousness*. Its meaning consists in the immediate expression of the very movement of material relations and it is this objective, *not conscious*, meaning that subjects first communicate to one another in their "material behavior." This meaning becomes subjective when the subject expresses it to himself, in the *internal language* or *consciousness*, which is thus originally a "*direct efflux of material behavior*."

The moment consciousness is constituted, we immediately find it implied in material behavior itself. In the dialectical flow of history, however, new meanings are always constituted which are at first unknown to consciousness in the language of real life, and only become the object of cognizance after a period of time. Consequently we are faced with two layers of meaning, conscious meanings and not yet conscious meanings which, nevertheless, have already been expressed in language.

The presence of a layer of non-conscious [pre-conscious] meanings in language has been demonstrated by Marx in *Capital* with reference to the "language of commodities":

> . . . everything our analysis of the value of commodities previously told us is repeated by the linen itself, as soon as it enters into association with another commodity, the coat. Only *it reveals its thoughts in a language with which it alone is familiar, the language of commodities*. In order to tell us that labour creates its own value in its abstract quality of being human labour, it says that the coat, in so far as it counts as its equal, i.e. is value, consists of the same labour as it [the linen] does itself.[20]

The "language of commodities" is the very language of the mercantilists insofar as it contains a layer of objective meaning, of which the mercantilists themselves are unaware: this is what Marx calls the "thoughts" not of the mercantilists but precisely of the commodity itself which *betrays* them through the language of trade and exchange. Thus when mercantilists say that a coat is equivalent to 20 yards of linen, and that it has consequently the same labor-value as the linen, they do not realize the true meaning of their speech, namely that the 'labor' of which they are talking is not concrete labor but the "abstract quality of being human labour." Classical political economy, which is only the "language of commodities" raised to a theoretical level, was never able to distinguish clearly between concrete and abstract labor.

Of course the distinction is made in practice, since labour is treated sometimes from its quantitative aspect, and at other times qualitatively. But it does not occur to the economists that a purely quantitative distinction between the kinds of labour presupposes their qualitative unity or equality, and therefore their reduction to abstract human labour.[21]

In other words, classical political economy expresses this distinction *without being aware of it*. This is what Marx particularly notices with regard to Franklin:

The famous Franklin . . . says this: "Trade in general being nothing else but the exchange of labour for labour, the value of all things is . . . most justly measured by labour" . . . Franklin is not aware that in measuring the value of everything 'in labour', he makes abstraction from any difference in the kinds of labour exchanged and thus reduces them all to equal human labour. *Yet he states this without knowing it*. He speaks first of 'the one labour', then of 'the other labour', and finally of 'labour', without further qualification, as the substance of the value of everything.[22]

The meaning of labor as abstract labor distinct from concrete labor is imposed by the very logic of the material relations in bourgeois society. Classical political economy could not avoid expressing it in its discourse when it describes precisely these relations, for the very movement of the discourse objectively expresses it. Nevertheless, the bourgeois economist does not become conscious of it. In fact, ". . . abstract universal labor . . . assumes the form of social labor *as a result of the universal alienation of the products of individual labor* . . . ."[23] The class interests of the bourgeoisie, of course, do not allow them to admit that its wealth, based on exchange value, has its origin in alienated labor. To distinguish clearly between abstract and concrete labor is to state the split between the capitalist and the worker, to recognize "*the universal alienation*" of workers in bourgeois society. That is why Franklin *could not speak to himself*; in other words he could not become conscious of the fact that labor as creator of exchange value is an abstract labor distinct from concrete labor, which creates use value. "Yet he states this without knowing it [*Was er nicht weiss, sagt er jedoch*]." He says it involuntarily, for this meaning is imposed objectively by the force of circumstances outside consciousness, in *the language of real life*.

If, however, we take up the general problem of meaning at the strictly human level, where consciousness is already established, the language of real life itself already clearly implies every conscious fact inherited from history. Consequently, meaning in its totality always seems to presuppose consciousness. We get the impression then of going in a circle: consciousness presupposes language and language, consciousness. We must begin by stating the problem starting with an *absolutely original* meaning, which appeared objectively in the language of real life before all consciousness in general, whose subjectification makes possible precisely the definition of the very first form of consciousness. This definition will be presupposed at all following stages as the previous condition of the use of language or thought. Such is precisely the meaning of the indicative gesture.

As the original sign of language, the indicative gesture exhibits this altogether singular distinctiveness, that through its simple material form, it produces its own meaning entirely by itself. By his circular-arc gesture the subject does indeed communicate to the other a movement in the form of an image which directs him toward the object: the gesture 'makes the image'. This image, which defines the very meaning of the gesture is precisely the projection of the real motion of the hand in a circular-arc upon the other subject. In fact, the indicative gesture as guidance at a distance, implies a *tendency* of the subject to go to the other in order to guide him toward the object. This tendency is of course a material one, since it consists, in the material form, of nerve synapses. But since it can actually only externalize itself from the arm's length to the end of the hand, it is by a *tendential extension* of the hand gesture that the subject reaches the other and directs him toward the object. This extension is really implied in the very tension of the arm and hand and it is immediately perceived within the framework of the sensori-motor psychism. Thus when the finger is pointed in order to indicate an object to an ape, his look follows the extension of the experimenter's hand gesture to the indicated object, and if this answers his need, he pounces upon it. "It even happens that when seeing the experimenter make a circular motion of the hand above the floor, the ape starts running around in a circle."[24] In the original form of the circular-arc gesture, the gesture of the subject is projected upon the other, and in this way draws the image of a movement going from this other to the object. Such an image is, of course, *non-material*. We may call it a *tendential image* since it consists of the tendential extension of the actual motion. Thus the original indicative gesture is a sign which, by means of its own material form, entirely produces its own meaning. This is not material since it consists only in a tendential image which clearly does not contain the least particle of matter. It is not ideal either, since as yet it does not imply any consciousness. Hence we shall call it a *tendential meaning*.

The notion of a tendential image can serve to characterize the sensori-motor psychism in general. When an animal sees an object, he sketches or outlines in his body an ensemble of movements which orient his behavior toward this object according to the tendencies already established by hereditary or acquired associations. It is these outlined movements which project the perceptive image of the object, which then guides the actual behavior of the subject.

This image is obviously not material. In fact, even though we say, generally speaking, that the subject has the image of the object 'in his head', this image is, strictly speaking, not found 'in' his brain, but rather *outside, exactly where he sees the object*. We have here a certain analogy with the phenomenon of the mirror. When we say that we see the image of the object 'in the mirror', this is obviously a manner of speaking, but one which is not at all exact, since that image is not 'in' but rather *behind* the mirror. It contains absolutely no trace of matter since it results simply from the virtual extension of the reflected rays, in such a way that everything happens for the observer as if they came from an object symmetrical to the real object in relation to the mirror.

In the case of the psychic image there is, however, an essential difference. In fact the projection, which constitutes this image starting from the outlined movements of the animal, is actually produced by the *tendency* of these movements. Consequently, in opposition to the virtual image of objects in the mirror, which is entirely unreal and which exists only for an outside observer, the psychic image has a *tendential* reality, so to speak, and it actually exists for the subject himself. It remains strictly *non-material*, however, since it does not contain any trace of matter either.

The tendential image of the sensori-motor psychism at the animal level does not, generally, have any meaning value, for the outlined movement which projects it, does not function as a sign. As we have already noted, the gestural activity of apes denotes feeling and action, and not 'meaning' strictly speaking, as the meaning of the object. It is only with the original indicative gesture that the psychic image becomes meaning. Indeed, the image of movement toward the object, projected by the gesture of the subject upon the other, thereby defines the *distance to be covered* or in other words the *relation of externality* between the other subject and the object. And as the gesture is reciprocal, the image returns to the first subject himself. This time, we are dealing with the meaning *of object*, even with its fundamental meaning, its externality to the subject.

As we have already noted, the animal does perceive the external object, but it does not perceive it *as external*. Now, owing precisely to the image projected by the original indicative gesture, the subject, receiver of the image, perceives the object *as external* for the first time. This image of the movement toward the object is completed on the sensori-motor image of the object itself, so that this image of the object is found in *the relation of the externality* of the object as related to the subject. In other words, it takes on the *meaning of externality*, a meaning which reflects the real externality of the object. We are still dealing here, however, only with a tendential meaning which, as such, does not stand apart from the gesture which projects it. The object is indeed perceived as external, but not yet as independent of the actual gesture of reciprocal indications. It is only through the transition to consciousness, where meaning will become *ideal* by detaching itself from the material reality of the signifying act, that the relation of externality will be disengaged as such. The object will then be perceived not only as external but also in its *objective* externality, as existing independently of the subject.

We have described earlier the genesis of the indicative gesture in a straight line, both as indication to oneself and as the predominant form of reciprocal indications. Once the form is constituted, the subject continues to see his own internal image in the enduring image of the others. "He sees and recognizes himself" in that multitude of calling images which reflect around him his own call to himself, and it is starting with that "*immanent environment*" that he sees himself pointing out the object to himself. We feel this very definitely within ourselves when we perform this gesture: it is always at a certain distance from ourselves that we address the indication to ourselves. As we have also noted earlier, it is the same for internal dialogue: it is from the position of another, which is at the same time myself, that I speak to myself in the second person, *as to another*. When this confusion between the signifying act of the subject with the image of the other reaches a certain degree of intensity, hallucination results, where the subject believes he hears voices or sees gestures which would be addressed to him by another person. Actually, *it is obviously he himself who is speaking to himself or addressing signs to himself by identifying himself with the image of this other*. These hallucinations, which actually belong to pathology, were absolutely common events in primitive societies, where they were considered normal, and where they provided the psychological basis for the imposition of religion as the dominant form of ideology. We find the same phenomenon, but in an attenuated form in the child who plays by himself, and it is altogether normal in the adult when he is dreaming.

Thus when the prehominid subject, left by himself, outlines to himself an indicative gesture of an object, he confuses himself with his own image in the stylized image of the others and recognizes himself there 'in himself'. It is this relation of *internal recognition* which defines the form of *lived experience* and the subject perceives himself in this form. In other words, he does indeed have the proprioceptive perception of the outlined movement of his hand stretching toward the object, but this perception is immediately absorbed in this immanent environment which reflects him from all sides, and his gesture appears to him as a simple moment in this set of images. Thus the subject's act of signifying the real, the material, takes the form of an *ideal indication* or *intentional* sighting in the *informality of lived experience*.

We have here something which is comparable, *mutatis mutandis*, to what occurs in those systems of multiple and parallel mirrors where the centrally placed subject sees his own image repeated indefinitely in all directions. Carried away by the general movement of these reflections of himself, he ends up feeling like a singular instance of this universality of images: an image reflecting all these images. "And if he happens to be in a mood to philosophize, he might happen to say, in a Berkeleyan style more or less tinged with Platonism, that this universal system of images, or 'ideas,' is the 'true Being' of which his own corporeal reality would be but the 'copy', the 'other-being' or 'alienated being'."

At the moment of the dawn of consciousness the idealization of the signifying gesture constitutes a decisive step forward which allows the detachment of the signified meaning from the material reality of the gesture itself. In other words, the tendential image of the movement toward the object, which defines the tendential meaning of the actual indicative gesture, becomes the ideal image, or ideal meaning, projected by the ideal or intentional act of the experienced pointing gesture. The *ideality of meaning* consists in its appearance of being simply carried by the experienced movement itself, in other words, by the movement of the reciprocal reflections in that 'immanent environment' just described, 'independently' of the material gesture which, henceforth, seems to figure only as a '*disappearing moment*'. And it is precisely in this way that the content of this meaning, that is, the movement toward the object, where the distance to be overcome is defined as the *relation of externality* between subject and object, is 'ideally presented' *in itself* independently of the actual motion of the subject's hand which projected the image of this relation. The meaning thus constituted immediately obscures the sensori-motor image of the object, which appears in its *objective externality*, as *independent of the subject*. In this way the transition is made from animal perception, as simple sensori-motor perception, to the original form of consciousness: *sense certainty*, as the call to oneself toward the object as an object of work, or the intentional sighting of the '*this here*'.

" ... Sensation," says Lenin, quoting Bogdanov, "is the foundation of mental life; it is its immediate connection with the external world. At each step in the process of sensation, a transformation of the energy of external excitation into the fact of consciousness takes place." [25] We know that every transformation of energy implies its conservation in a form different from the motion of matter. The fact of consciousness in which the energy of the external excitation has been transformed, thus necessarily includes a material motion. Naturally, this is not just a simple physico-chemical or biological movement. The existence of matter at the human level takes the form of society. And the completion of the material motion in its social form, that in which society is presented as such, its *being there* is language. "Language itself is the product of a community, just as it is in another respect itself the presence [*Dasein*] of the community, a presence which goes without saying." [26] The material motion implied in consciousness is thus language itself as the material motion of the signifying act or linguistic sign.

The linguistic sign, naturally, can only be considered as a constitutive moment of consciousness if it somehow implies its own meaning. If it is presented merely as an arbitrary sign, it will only be a simple external expression, all meaning being derived from consciousness. On these conditions, the fact that it is always tied to language will effect its nature as such. In other words, consciousness will always be defined in itself as purely lived internality which makes its relation to matter inexplicable.

We have shown earlier how the original linguistic sign, the indicative gesture in a circular-arc form, produces its own meaning entirely by itself in the form of a tendential image of a motion going from the other subject to the object. The notion of *tendential meaning* can serve to characterize generally the language of real life insofar as it directly reflects the motion of material activity and the material relations of men independently of their consciousness. Thus, in a passage quoted earlier, in Franklin's statement commented on by Marx, the distinction between concrete and abstract labor appears in the very juxtaposition of the two propositions. In the first, the word 'labor' used twice, is necessarily understood in the concrete sense, since we are dealing with an exchange of labor for labor. On the other hand, in the second proposition, the same word occurs once, in a general sense. In the dichotomous structure of the sentence there is thus an *objective tendency* to distinguish two meanings of labor, even though the author is not aware of it, since he expresses them by the same word.

This disposition of the verbal ensemble is imposed by the very nature of things, namely the dialectical opposition between use-value and value in the material activity of the exchange of commodities, which implies the opposition of the kinds of labor which have created them. And it is in this way that this real opposition has been reflected, unknown to the author, in the *tendential meaning* of his discourse which, while presented as a perfectly conscious theoretical exposition, nevertheless includes the non-conscious moment of the *language of real life*.

We thus see that at this level, meaning, at least in its new layer, is directly produced by the material motion of the signs themselves, insofar as it is necessarily shaped by the motion of things. Naturally, if we take that meaning in its totality, it also implies the whole of the already acquired content of consciousness. But since this total content has itself been established historically on previous forms of the language of real life, where it was presented only as tendential meaning, we see that it is possible, starting with the original indicative gesture and proceeding gradually following the dialectic of history, to demonstrate that the ensemble of meanings that we presently possess, has arisen from the material motion of linguistic signs in the language of real life "or, in a more general way, in social practice."

Every meaning, however, once it has become conscious can, in principle, be tied to any verbal sign whatsoever. The word *arbre* can be expressed just as well by the words *arbor, tree* or *Baum*. The relation of the signifying to the signified appears to be a matter of purely arbitrary convention. In reality, however, this is just a specific characteristic of the verbal sign which we know does not at all exhaust the material reality of the signifying act. This act always includes gestures, which by themselves produce their own tendential meaning. The gesture 'makes the image', and in the course of history, it becomes more and more representative by taking the operative form of schema, drawing, etc.

Clearly, when we pronounce a word in inner speech, the outlined gesture which accompanies it is not sufficient to determine its meaning as it appears to consciousness. But it is appropriate to point out that the verbal sign is itself already associated with an ensemble of gestural movements. These are *evoked* by utterance in such a way that as the gestural component in the material motion of the signifying act of inner speech we have not only an actually outlined gesture, but what is more, by virtue of an *evoked outline*, the entire operative system, which allows the definition of the meaning of the verbal sign.

In fact, it is clear that the evoked outline is as real, as material, as the actual outline. We know that it is possible, by picking up and amplifying the bioelectric currents in the forearm, to obtain movements in an artificial hand which result from the subject's mental activity, his own hand remaining motionless. Scientists are presently thinking of drawing these currents directly from the motor zone of the brain rather than from the effector organs. The evoked outline is thus distinguished from the actual outline only to the extent that the latter is composed of a slight muscular movement, effective though not externalized, while the former is limited to the nervous system. Thus we see that in inner language, the material motion of the signifying act taken in its totality, with its operative gestures actually outlined or simply evoked now enables us to define its meaning directly as the *tendential image* projected by these same gestures. On the other hand, as we have seen earlier, the subject can address himself only through the stylized image of others, where "he sees and recognizes himself." The signifying act is thus immediately sent back to himself in that 'immanent environment', in other words, it is experienced as an ideal act, where its tendential meaning stands out from the material reality of the gestures which have projected it, and it becomes ideal meaning.

We thus see that the *ideality* of consciousness is not some kind of ideality 'in itself', but is constituted in the actual motion of *idealization* which is immediately implied in inner language. Such an idealization cannot, of course, suppress its material foundation, particularly the real gesture which is impressed in the ideal motion itself. This can be verified in the analyses of the idealist philosophers themselves. "We cannot think a line," says Kant, "without *drawing* it in thought, or a circle without *describing* it. We cannot represent the three dimensions of space save by *setting* three lines at right angles to one another from the same point."[27] These are manifestly completely real operative gestures which Kant considered abstractly in their pure idealized form, as pure mental operations accomplished in some kind of thought in itself.

Obviously, when we examine the epistemological relation of knowledge to its object, consciousness must be considered in the very form in which it enters that relation, as ideal image of the external world. *From this point of view*, the opposition between matter and consciousness has an absolute meaning. But it is precisely only to stamp, in an absolute way, the nature of consciousness as *secondary*, insofar as it is a reflection, or image, in relation to the matter of which it is the image. The opposition between the ideal form of consciousness and the materiality of its object thus does not exclude, but implies, the dependency of consciousness, as a secondary factor in relation to matter, which is the primary element. And this dependency obliges us precisely to conceive of consciousness as a product of matter. The opposition between the two terms thus has in itself only a *relative* value: " . . . the antithesis of matter and mind," says Lenin,

has absolute significance only within the bounds of a very limited field – in this case exclusively within the bounds of the fundamental epistemological problem of what is to be regarded as primary and what as secondary. Beyond these bounds the *relative character of this antithesis is indubitable.*[28]

The relativity of the opposition obviously consists in the exclusion of dualism and implies the *fundamental unity* of the two opposed terms, a unity which is expressed in *material monism*. "The materialist elimination of the 'dualism of mind and body' (i.e., materialist monism)," says Lenin, "consists in the assertion that the mind does not exist independently of the body, that mind is secondary, a function of the brain, a reflection of the external world."[29]

Naturally, we must not conclude from this that consciousness would itself be material. For the assimilation of consciousness to matter would end up in pure identity, which could just as well result in the inverse assimilation of matter to consciousness. " . . . to say that thought is material is to make a false step, a step towards confusing materialism and idealism."[30] It nevertheless remains true that " . . . thought and matter are 'real', i.e., exist . . . "[31] And from the point of view of materialistic monism, one must indeed admit that consciousness, insofar as it actually exists must have something material in itself. "And since its form is already ideal, the material aspect implied in it can only be its substance, matter *qua subject*. This substance or this subject, implied in consciousness, is exhibited first as the matter of the linguistic sign, as *'social matter'*."

Social matter, in its linguistic layer, is defined as the ensemble of signifying acts, gestures and utterances, *in the structure of reciprocity*. In the course of history, this structure is exhibited in an infinite variety of forms and in ever increasing complexity, but all of them are based upon a fundamental form of *repetition*, or *echo*, which we have described as the indicative gesture. Thus the individual who speaks and makes a sign to himself in inner language, receives an immediate replica of himself in the stylized enduring image of others, projected first by virtue of the tendential image, by the motion of the cerebral traces left by the social experience of reciprocity in which he constantly sees his own gestures as in a mirror, and hears his own voice as in an echo. In the historical development of human societies, particularly in the dialectic of class relations, the movement of reciprocity is laden with a content of differences and contradictions which are reproduced naturally in an echo form from the 'immanent environment' and diversify this image of himself that each finds in the stylized enduring image of the others. The subject addresses himself starting essentially with the image of his own social group but including other groups at the same time, and he has the echo of his inner language in various forms in the stylized image of the various groups of his social environment in general. He recognizes himself in the image of the others in a form that is *identified, modified*, and *oppositional* or *antagonistic* according to whether it concerns his own group or different groups, friends or enemies. Lived experience is thus constituted as a *singular dialectical unity of differences and contradictions* which express the real activity of social differentiations and contradictions in the subjectivity of individual consciousness including the various possible errors and confusions.

In short, the individual can address himself only insofar as he is a social being, as having already received in himself the form of social reciprocity. "It is clear . . . ," says Marx, "that [the individual] relates even to language itself *as his own* only as the natural member of a human community." [32] Thus it is social matter indeed, as *linguistic matter*, which, by means of its reciprocal form embedded in the cerebral traces of the individual, *produces* this movement of idealization, in which the original tendential relation of the signifying to the signified, is sent back to itself, and thereby becomes the lived relation of the ideal act of consciousness to its intentional meaning. "Consciousness is, therefore, from the very beginning a social product . . . " [33] This product has the following characteristics: on the one hand it *stands out* from the material movement which produces it – so that it cannot be considered material itself – but on the other hand it cannot be *separated* from that material movement from which it stands out, since it exists only in that idealization process which always begins at each instant of the reciprocal form of the linguistic sign itself. In other words, the ideal movement of the act of consciousness stands out from the material signifying act as a figure 'stands out' from its background, without being separated from it, since it is the very presence of that background which enables the figure to stand out from it. The movement of consciousness, produced by linguistic matter, therefore does not exist apart, in itself; in other words, it continues to belong to that matter.

"Motion," says Engels, "in the most general sense, conceived as the mode of existence, the inherent attribute, of matter, comprehends all changes and processes occurring in the universe, from mere change of place right to thinking."[34] The motion of thought or of consciousness, being thus included in the general motion of matter, can itself be only movement of matter. More exactly, Engels continues, " . . . the motion of matter is not merely crude mechanical motion, mere change of place, it is heat and light, electric and magnetic stress, chemical combination and dissociation, life and, finally, consciousness."[35] In the *Preliminary Notes* to *Anti-Dühring* Engels also says "motion in cosmic space, mechanical motion of smaller masses on the various celestial bodies, the motion of molecules as heat or as electrical or magnetic currents, chemical decomposition and combination, organic life up to its supreme product, thought – at each given moment each individual atom of matter in the world is in one or other of these forms of motion, or in several forms of them at once."[36]

These formulae obviously do not in any way diminish the specific, ideal character of consciousness. To say that consciousness is a motion of matter simply means that this motion does *belong to matter*, in other words, that matter is its true *subject*. It nevertheless remains that this motion is exhibited not in material, but in ideal form since it results precisely from the *idealization* characteristic of the motion of the inner language.

Consequently, we see that the fact that consciousness belongs to matter does not exclude but implies its form of ideality, just as that form, as the idealized form of the motion of inner language, necessarily implies the presence of matter as *subject*. Every study of consciousness must therefore consider it in its double aspect, on the one hand in its *ideal motion* as image of the external world, and on the other hand in its actual reality as ideal motion *of matter*, in other words, of the real social man and more precisely of his brain. This is the reason why Lenin defines the object of the theory of knowledge, i.e., the relations of knowledge, not by two but by three components. "Here there are *actually*, objectively, *three* members: (1) nature; (2) human cognition = the human **brain** (as the highest product of this same nature), and (3) the form of reflection of nature in human cognition, and this form consists precisely of concepts, laws, categories."[37] It is clear that it is in the consideration of that third component, "the form of reflection", that is, the idealized form in which knowledge is exhibited as ideal image of the external world, that the antithesis between consciousness and matter appears absolute. Thereby is resolved "the fundamental epistemological question, what is primary and what is secondary?" For it is rather obvious that the ideal image of the external world can only be secondary in relation to its material reality. But outside of "the very limited bounds" of that question, epistemology must also consider knowledge in its effective reality, as the concrete movement of its historical becoming. Lenin here does not hesitate to place the equal sign between human knowledge and the human brain. Knowledge is *the brain itself in its motion of thinking*: "matter *which* thinks" or as Engels liked to repeat, "the thinking human brain."[38] And in this motion of the human brain, as the superior product of nature which ideally reproduces this same nature within itself, it is, in the final analysis, *nature itself which knows itself.* Man, says Engels, is " . . . that mammal in which nature attains consciousness."[39]

Consciousness is thus truly a motion of matter and matter is the true *subject*. There always arises a certain confusion, however, when consciousness is defined as a "form of the motion of matter" and we may notice that Engels was particularly careful to avoid such a formula, which very naturally brings to mind a simple material motion. Consciousness must, therefore, be more exactly defined as the *idealized form of the motion of inner language*. And since it exists only in that language, the only "form of the motion of matter" in question here is, strictly speaking, language itself. Language first of all objectively consists of material behavior as the language of real life, a direct expression of material activity and of the material relations among workers, and *raises itself to consciousness* in inner language where the subject addresses himself starting with the image of the others in which he recognizes himself in the identity of his own lived experience. The *intimacy of consciousness of the internality of lived experience* is this idealized form where the subject assimilates social experience sanctioned by language for himself, and relates it to himself by confusing himself with this 'internal society' where at each moment – in a real or illusory manner – he finds his *form of universality as the mode of his existence*. "Man," says Marx, "is in the most literal sense of the word a *zôon politikon* , not only a social animal, but an animal which can develop into an individual only in society."[40]

SECOND INVESTIGATION

SYNCRETIC LANGUAGE

INTRODUCTION

Roman Jakobson, following a thesis of C. S. Peirce, says, " . . . any sign translates itself into another sign in which it is more fully developed." Thus the meaning of a word will be given by its dictionary definition, its translation in another language, its pictorial representation, etc. "But in all these cases we substitute signs for signs. Then what about a direct relation between sign and thing?"[1]

It is quite clear that if the whole meaning of signs merely consists in their referring to one another, without ever referring directly to things, then we are practically enclosed in a world of signs, so that we no longer see what speaking of things could mean. But then, in effect, it is the very existence of things which becomes questionable: "As symbolic logic has persistently reminded us, 'linguistic meanings', constituted by the pattern of analytic relationship of one expression to other expressions, do not presuppose presented things."[2]

In other words, from the simple definition of the concept of meaning, presented within the framework of the science of language, one easily passes to a philosophical position which, if not denying the reality of the external world, at least declares it to be useless and void of meaning.

On the very level of sign analysis, however, it is altogether impossible to ignore the existence of a fundamental sign, whose meaning consists precisely in positing "a direct relation between the word and the thing": the indicative gesture which points the finger to *the thing itself*. The simple indication of the object, it is true, does not yet give the receiver of that sign any actual information about the particular properties of the object indicated, and this is what Jakobson has not failed to point out.

> Suppose I want to explain to a unilingual Indian what Chesterfield is and I point to a package of cigarettes. What can the Indian conclude? . . . He will gather what Chesterfield is and what it is not only if he masters a series of other linguistic signs, which will serve as interpretant of the sign under discussion.[3]

In other words, the meaning of the indicative gesture is to be found only in other signs so that we are always dealing with signs which 'interpret' one another, and never with things.

This is obviously a grave error.

For once the package of cigarettes in question is pointed out to the Indian, all of the other signs that we may add to it will function not as 'interpreting' the indicative gesture, but as means to exhibit the particular properties of the indicated object, which is altogether different. The indicative gesture simply means that it is a question of this very object, the *'this here'* as objective reality given to sense intuition and *nothing more*. Such a meaning is understood by the gesture itself, it has no need to be interpreted. Now if, for example, one adds by a mimetic sign that it is something *to be smoked*, one will have shown a certain particular property of that object, and will not have explained the meaning of the gesture of pointing with the finger. For it is quite clear that the fact of pointing to a thing does not mean that it is supposed to be smoked.

In other words, the word *'Chesterfield'* has a complex meaning, of which only one component is explained by the indicative gesture. This component is, to be sure, a fundamental one since it concerns the objective reality given to sense intuition. No matter how poor such a semantic content may seem, it is, nevertheless, an altogether essential one, since it allows one to distinguish the meaning of the word *'Chesterfield'* from, for example, the word *'chimera'* which downright excludes the possibility of pointing at the object with the finger. It goes without saying, however, that once the unilingual Indian has learned that Chesterfields belong to the class of real objects as opposed to imaginary objects, he still will not be able to distinguish them within that class. He will be able to do so "only if he masters a series of other linguistic signs which will serve as interpretant" not, of course, the indicative gesture, but the word *'Chesterfield'* itself insofar as it contains semantic components other than the one explicated by the gesture of pointing at the object with the finger.

In short, the meaning of the indicative gesture in no way refers to any other sign. It uniquely and directly refers to *the thing itself* in its external existence as independent of the subject, in other words in its *material* existence. "Matter," says Lenin, "is the objective reality, which is given to us in sensation."[4] Naturally, the word "given" must be understood here in an active sense. "The senses," specifies Lenin, "*show* reality."[5] They show it to us precisely because of the indicative gesture which defines the very *act* of sense intuition as *sense certainty*. Objective reality is *'given'* to us in sensation in the sense that it is *indicated* to us by sensation. And since sense intuition is also implied in conceptual knowledge as its necessary foundation, it is, consequently, the total object of knowledge in its singular reality and general determinations which we *show* or *indicate* to ourselves: "The senses show reality, thoughts and words, generalities."

We have already shown how the indicative sign, which appears at the very origin of consciousness, effects the fundamental mediation between social practice and lived knowledge, a mediation which assures the correspondence between knowledge and things. It is the meaning of this sign which is the basis of the concept of *matter*, as an essential concept of the theory of knowledge. This meaning does indeed consist in the image projected by the finger stretched toward the object, an image which reflects the real relation of externality, in which the object is presented as external reality independent of the subject. Now it is this very determination of external existence which defines the philosophical content of the concept of matter. "The *sole* 'property' of matter with whose recognition philosophical materialism is bound up is the property of *being an objective reality*, of existing outside the mind."[6] We see that, in the final analysis, the concept of matter refers back to the indicative gesture as it constantly arises from *human social* practice at its deepest layer, and is thus constantly implied in the very function of our sense organs. "Acceptance or rejection of the concept [of] matter," says Lenin, "is a question of the confidence man places in the evidence [*pokasaniiam*] of his sense-organs, a question of the source of our knowledge . . . "[7]

Quite obviously the simple presentation of the object as *matter* does not yet give us actual knowledge of it. Matter only exists in motion and "the world is the movement of this objective reality reflected by our consciousness."[8] Knowledge will then be actual only to the extent that it will show us the object as *matter in motion* " . . . [so that] the life of the subject-matter is now reflected back in the ideas."[9] In other words, the indicative sign must be developed by *following the motion of the object*. Thereby the motion of knowledge is constituted as the reproduction or as the more and more approximate image of the real motion of things. "With me," says Marx, "the reverse is true: the ideal is nothing but the material world reflected in the mind of man, and translated into forms of thought."[10]

I. THE DEVELOPMENT OF THE INSTRUMENT

From Prehominid to Homo Habilis

We have shown how the indicative gesture came into being in adaptive collective work as it developed beginning with the anthropoid stage. With the transition from the anthropoid ancestor to prehominid (Australopithecus or Australanthropus) at about the end of the Tertiary or Quaternary,[11] we may consider the form of the sign to have become an acquired form of behavior. It was indeed at this time that the foot became specialized in bipedal gait, which freed the hand to become accustomed to holding instruments. Consequently, adaptive work which, at the end of the anthropoid ancestor's development, was still merely simple habit, became for the prehominid a regular behavior based on the progressive development of the biological structure.[12] Now, such a development presupposes that the indicative sign, which was implied in the most elementary form of the collective use of instruments, became established in the nervous system as an available form of behavior.

It is probably the exercise of labor, at the prehominid stage, that produced the cognizance of this first linguistic sign at the beginning of this stage, which constituted the original form of consciousness as *sense certainty*. The acquisition of sense certainty, in its turn, gives a new impulse to adaptive work, the development of which involves the earliest progress in the development of language and consciousness. In such a movement of reciprocal action, adaptive work develops into increasingly complex structures and finally ends, with the appearance of *Homo habilis*, in the first form of productive labor. The cultural remains discovered in Bed I of the Olduvai Gorge show in fact that the inhabitants of that camp, in the upper Villefranchian,[13] already possessed a rudimentary stone-cutting technique. On the level of consciousness, such an activity implied the presence of an ideal image representing the typical shape to be imposed upon the material. "At the end of every labour process, a result emerges which had already been conceived by the worker at the beginning, hence already existed ideally."[14]

It is true that the attribute of 'tool' given to the Olduvai stones has been the subject of serious reservations. Their atypical and even 'amorphous' character has been emphasized. We are dealing here with stones that have been worked on their useful part only, generally consisting of a sharp or pointed edge produced by five to eight rough cutting strokes. The rest has retained its natural accidental shape so that each specimen appears to be different. The work of shaping does not seem to have been guided by the ideal representation of a typical model. The 'production' of such 'tools' would not go beyond the framework of reflex conditioning and the so-called '*Homo habilis*' would only be a particularly evolved Australopithecus.

In his 1966 report on the Olduvai culture, Leakey, however, assures us that " . . . there is a considerable diversity of tool types and close conformity within certain groups of tools, indicating that the tool-making had already attained a measure of standardization, even at the period represented by the lower levels of Bed I . . . "[15] The inventory that is given us is, in fact, rather impressive: choppers, spheroids, proto-bifaces, scrapers, proto-burins, anvils, various flakes. To which it must be added that within the category of choppers there are up to seven distinct varieties.

We believe that this is a misunderstanding. The Olduvian implements were certainly typical *in their useful part*, since this was produced by work and can serve as a basis for classification. But their *general shape* remains ill-determined, for the part that is not worked on is by far the largest. Taken as a whole, this 'tool' gives the impression of a stone that has been more or less 'contrived' as such. One could thus speak only of a *partial typical shape*, or if one wishes, an *embryonic* shape. In fact, the useful part which bears the human mark of the productive act, still remains, so to speak, embedded in the ensemble of the natural shape of the stone, somewhat like the embryo in the maternal womb. Such a situation necessarily contains a certain equivocation: "though the embryo," remarks Hegel, "is indeed *in itself* a human being, it is not so for itself [*für sich*]."[16] We arrive at the same result if we examine the morphological data. From anthropoid to prehominid and from prehominid to *Homo habilis* we see the progressive refinement of the jaw and the growth of the brain which seems to indicate the first development of language and consciousness. With a height that does not exceed that of the chimpanzee (1.50m), the Australanthropus already possesses a brain somewhat larger than the gorilla's: 508 cm^3 versus 498 cm^3 for the gorilla and 393 cm^3 for the chimpanzee (Tobias). *Homo habilis*, who is only as tall as a pigmy (1.20 m), already possesses a brain of 657 cm^3 (Tobias, 1968) which represents an even more important advance over Australanthropus than that of Australanthropus over anthropoid. It seems, therefore, that the possibilities of animal organization have been largely by-passed. And yet, *Homo habilis* has not yet crossed the famous 'cerebral Rubicon' which, according to many writers led to humanity, and which is placed between 700 cm^3 (Weidenreich) and 800 cm^3 (Vallois). It must be added that the admission of this new species in the genus *Homo* was only made possible at the price of a serious revision of the definition of the genus.[17]

The problem therefore remains controversial.

Since it appears to be difficult to resolve this problem from mere examination of the Olduvian stones, or from morphological comparison among hominid fossils, we believe that one could try approaching it from the preparation of instruments by the ape. There does indeed exist a precise criterion for the objective definition of the limit of this operation and, thereby, a presentation of the conditions for its transcendence.

From the Preparation of the Instrument to its Elaboration

When an anthropoid makes itself a stick by breaking off twigs from a branch, by unbending a coiled wire, by tearing up a piece of wood in order to form it into a stick, or by fitting bamboo reeds together to make a stick long enough to reach a distant object, he *always uses his natural organs exclusively*, that is, his hands, feet and teeth, *without ever using an intermediary object as an instrument of labor*. In experiments on this subject, performed by Khroustov, in particular, it was never possible to train the ape to make one implement by means of another.[18] We are faced here with the limitations of animal intelligence, the transcendence of which must mark a fundamental stage on the way to tool production.

As a matter of fact the preparation of the instrument by the ape cannot even be considered, strictly speaking, as an act of work. "The simple elements of the labour process," says Marx, "are (1) purposeful activity, that is work itself, (2) the object on which that work is performed, and (3) the instruments of that work."[19] Since the anthropoid prepares the instrument without ever inserting any means of labor between its own natural organs and the material, this operation by itself does not constitute an act of work, but a simple act of direct manipulation. Work takes place only when the ape utilizes the instrument for the satisfaction of its needs. Consequently, in the whole process only the *object of need* functions as an object of work. The material employed by the anthropoid for adaptation to the situation *does not function as a work-object*, since it is directly manipulated by the subject without the intermediary use of any instrument of labor whatsoever.

Concerning the instrumental activity of the ape, Wallon remarks that "between the subject and the object, the desire to appropriate the object creates a field of forces the possible configurations of which depend at the same time on the vicissitudes of the external field and the circuits which can open up in the nervous system .... Closely complementary, the two fields are barely distinguishable, except for the analysis of their respective conditions."[20] In other words, in the total dynamic field, in which the *situation* of the perceiving subject is defined, the object of need attracts another object, which, owing to previous conditioning, appears as a potential intermediary object. *The object of need, mediated by this instrument, becomes the work-object*. Once the anthropoid is accustomed to the use of instruments, he can begin to prepare them if he does not find them ready-made in a situation which calls for their use. But since, at the animal level, need is defined as essentially biological, it is clear that the preparation of instruments is brought about uniquely by the object of biological need: in itself it does not answer to any need. Consequently, the material, indifferent in itself, is incapable of attracting a third object as an instrument of labor.

It is true that, starting from a certain degree of development, because of the diffusion of biological need, a sort of immediate need to prepare the instrument is produced. Thus the chimpanzees that are already accustomed to the preparation of instruments prove capable of attentive and persevering efforts, for example, in order to bite off a piece of wood to use it as a stick. The raw material appears here somehow as the object a new need: *the need to transform it into an instrument*. But it is quite clear that this need is only brought about by the presence of the biological object and is entirely subordinated to it. It is thus always the object of biological need which plays the role of the major center of attention and, so to speak, the dominant pole of the total dynamic field of perception. In this way the raw material appears here only as a subordinated pole, by virtue of which it cannot yet bring about the intervention of a second instrument. In other words, it continues to function only as an object for direct manipulation.

The use of a second instrument to act on the raw material is only possible at the moment when the preparation of the instrument occurs *in the absence of the biological object*, which is replaced by its simple *representation*. In fact, in such a situation, the raw material as object of the need to transform it into an instrument, becomes dominant in the dynamic field of perception, which allows it to attract a third object as an intermediary. The raw material then functions as a work-object, and the preparation of the instrument is raised to the level of an act of work.

Such a progress presupposes, as we have just noted, that the subject is capable of representing to himself the absent biological object, for the instrument, obviously, cannot be prepared for its own sake. At this level, the need to prepare the instrument can emerge only from the diffusion of biological need so that this preparation must be motivated by the presence of the object of biological need, if not in flesh and blood, then at least in its image.

We have seen that consciousness appeared at the beginning of prehominid development by the subjectification of the indicative sign in the form of the *sense certainty* of the objective reality of the object perceived.[21] From there to the *representation* of the absent object, one must obviously pass through a long mediation of work and language. We may thus believe that in a first stage of prehominid development, the preparation of the instrument was only accomplished in front of or beside the biological object, as we can observe in anthropoids, or at least in this immediate proximity, so that it was *still* more or less *present* by its *enduring image* in the perceptive field. Under these conditions, the biological object always functioned as the dominant pole of the situation, and the insertion of a third object as intermediary between the subject and the raw material was not yet possible. Results of excavations in presently known Australanthropi camps have exactly confirmed this, since until now we have been unable to find any object that could serve as an instrument which bore the mark of shaping by an intermediary instrument.[22]

We can thus consider the presently known Australanthropi as late representatives of the *first stage of prehominid development*. It is precisely in the course of this first phase that, through the development of work and language, conditions arose which finally made possible the appearance of a first representation of the absent object. Once this progress of consciousness was attained, the prehominid ancestor could prepare the instrument *far from the biological object*, at the start of a hunt, for example, since he already had in mind the image of the game. The raw material, under these conditions, comes to the foreground as the dominant pole of the perceptive field, and can therefore attract a third object to function as a second instrument.

The intervention of a second instrument transforms the direct manipulation of the raw material into an act of work, but of course we are not yet dealing with labor of production. The act of production implies that the worker guides himself by the ideal image of a *typical* shape, since it is precisely the presence of such a shape in an object which enables it to be recognized as the *product* of a human hand. At the point at which we have arrived, nothing authorizes us to assume that the subject already had such an image, whose level is clearly superior to that of the simple representation of an absent object. It is thus probable that the ancestor contented himself with shaping the raw material more or less approximately, as long as he obtained from it a more or less usable shape. We can speak here of a *work of elaboration*: the elaborated instrument must bear the mark of shaping by the intermediary of another instrument but still lacking a typical shape. Such a characteristic corresponds perfectly with Engels's indications, in the passage cited earlier, of the very limited character of the progress of labor in the intermediary or transition stage from ape to man: "At first, therefore, the operations, for which our ancestors gradually learned to adapt their hands during the many thousands of years of transition from ape to man, could only have been very simple. The lowest savages, even those in whom a regression to a more animal-like condition, with a simultaneous physical degeneration, can be assumed to have occurred, are nevertheless far superior to these transitional beings." Since there is no production as yet, we still remain within the prehominid framework, but at a higher level than the first phase mentioned earlier. The intervention of a second instrument, which transforms the direct manipulation of raw material into a work of elaboration, therefore marks the transition to a *second stage of prehominid development.*

The Genesis of Stone Working – The Kafuan as the Second Stage of Prehominid Development

Up to now, we have not yet found any bones that would allow us to picture concretely the Australanthropi who evolved from the second stage of prehominid development. Prehistoric pieces, however, classified as '*Kafuan*' [after the river Kafue in southern Africa], can serve as proof of their existence. These are stones having an irregular edge produced by one or two cutting strokes on just one side. The absence of any typical shape, even on the useful side, makes these pieces difficult to distinguish from natural instruments, and some authors attribute a purely physical origin to them. Nevertheless they do correspond to a necessary stage in the evolution of stone implements, which we can reconstruct from the simian level.

The usage of stone as an instrument is found in Capuchin monkeys who use stones to crack nuts, and in Japanese Macaques which use them to break crab shells.[23] We may believe that toward the end of his development the anthropoid ancestor got into the habit of using sharp stones for cutting up game. Such a habit led him to prepare these instruments when he could not find them ready-made in nature. The manipulation of stone, however, as opposed to that of wood, presents a particular difficulty: it is not possible to break a stone with the hands. It can, of course, be notched or chipped with the teeth, but the procedure is painful and inefficient. It is probable that the anthropoid ancestor and consequently the Australanthropus of the first prehominid stage had recourse to *breakage*. This could still be observed in the last century among the Tasmanians who broke a stone by throwing it against a rock, or on another stone, and then searched among the broken stones for those that could be used as sharp instruments.[24] Such a procedure does not go beyond the limits of a direct manipulation of raw material, for there isn't as yet any *instrument of labor*, since the second stone does not function as intermediary between the subject and the object of his action.[25]

At a certain stage of development, breakage can take the form of *crushing*: the subject holds one stone in each hand and strikes them one against the other. We find ourselves here at the threshold of stone-cutting. However, as the useful effect can be obtained just as well on one stone as on the other, it is clear that they both serve as raw material, so that we always remain within the framework of a simple act of direct manipulation.

As long as Australanthropus's preparation of instruments was restricted to the presence of objects of biological need, he could not go beyond the method of crushing. With the object of biological need monopolizing the dominant pole of the dynamic field of perception, neither of the two stones could become the principal center of attraction, so that their roles could not be differentiated into instrument of labor and object of labor. The situation changes from the moment the subject avails himself of the representation of the absent object. Once the game has been killed by means of a stick, for example, some hunters could, by keeping that image in mind, go rather far to search for and, if necessary, prepare a sharp stone with which to cut up the game. The object of biological need is pushed to the background of the total dynamic field and the two stones to be crushed both occupy the dominant pole. Now, if one happens to be chipped, but in such a manner that it is not yet usable, it can become the center of attention for the rest of the operation. In other words, it alone can become the dominant pole of the perceptive field. The other stone then takes on the function of instrument, the first one becoming, thereby, the object of labor. With the accumulation of experiences of this kind, the role differentiation of the two stones becomes habitual, in other words, from now on, from the beginning of the operation the subject takes one of them as an object of labor and the other as an instrument of labor: *crushing has become stone cutting*, and the preparation of instruments has been raised to the form of a *labor-process*.

The transformation of raw material by work evidently implies a certain representation of the useful shape to be made. Precisely there lies the superiority of the elaboration of the instrument over its simple preparation, such as we can observe among the anthropoids. The instrument that the ape manages to prepare is of any shape whatever as long as it can serve as an intermediary to obtain the coveted object. In the process of preparation, the subject is thus guided not by a representation of the useful shape of the instrument or instrumental form but simply by a 'generalized image'[26] resulting from his previous sensori-motor experience, whose content is essentially defined by a certain useful motion characteristic of the *instrumental function*: for example, in the case of the stick, the motion to reach the distant object. Such an image guides the ape in the preparation of his instrument applying itself concretely to the perceptive image of the present object of biological need, and we immediately see that the instrument so prepared is determined solely on the *functional, non-formal* plane.

We may believe that, in the first phase of his development, when the prehominid prepared the sharp stone by crushing, he was also guided by a simple functional image of that instrument, namely the generalized image of the motion of cutting, immediately applied, for example, to the game just killed, and which remains, more or less present in the dynamic field of perception. But from the moment that the ancestor set about to search far away for the necessary stones, the representation of the biological object, which was sufficient to motivate such a search, could not furnish a focus of attention consistently, firm enough for the functional image of the instrument to be prepared. As a result, the latter became too vague to be able to guide the preparation in question continuously. At crucial moments, the subject therefore had progressively to constitute a certain representation of the *instrumental shape*, or the cutting shape of the instrument itself as it appears in the motion of cutting, and it is precisely such a representation which enabled him to progress from crushing to cutting. In fact, the differentiation of the two stones, one becoming an instrument of labor, the other an object of labor, implies that the subject tries to guide the instrument of labor in such a manner as to produce a certain shape in the work-object: the more or less tapered shape which makes cutting possible. A representation of the tapered cutting shape was thus presupposed.

At this stage, obviously, there cannot yet be a question of a typical representation. The represented image of the instrumental shape, as it was constituted on the basis of the previous practice of work, could not yet be clearly disengaged from the accidental traits belonging to the natural instrument or the simply prepared instrument with which it was completely confused. The emergence of the shape, which would allow its function as a typical model, will be possible only after a long evolution of the work of elaboration, where the instrument will finally acquire a distinct shape at least in its useful part. At the present stage of development, the moment when the work of elaboration is only beginning, the represented image of the instrumental shape necessarily remains to a certain extent mixed with the contingent traits of the raw material: by virtue of that more or less confused mixture we may call it a *syncretic image*.

Now it is precisely a shape of this same kind that we find among the Kafuan stones. Their useful part does indeed consist in a roughly cut edge on *one side only*, so that the cutting instrument is necessarily irregular since it is made up of a natural side and a worked-upon side. Such a shape is defined as *syncretic* since it implies to a certain extent a more or less confused mixture of the strictly useful traits of the instrumental device and of the contingent traits of the natural side. And we can infer from this syncretic character of the Kafuan instrument that the labor-process which elaborated it was guided precisely by a syncretic image.

Of course, such a characteristic makes the Kafuan stones difficult to distinguish individually from natural instruments. If, however, we take them as a whole, *in situ*, we see an evolution which tends to link them to the Olduvian type:[27] yet such an evolution would be inexplicable in the hypothesis of a purely physical origin.

From the Elaboration of the Instrument to its Production – The Olduvian as the Final Stage of the Gestation Period

As long as the edge of the stone is worked on one side only, it necessarily retains its syncretic shape. Typical shaping will be possible only when the edge is cut on both sides and this is precisely what we can establish in the case of the Olduvian chopper. The transition from the *Kafuan* to the *Olduvian* thus implies as a prior condition the transition from cutting the stone on one side only to cutting on both sides,[28] which gives the useful part of the instrument a *distinct* shape. More generally speaking, it was first a question of the transition from a *summary*, or *confused*, shaping or *semi-shaping*, to a *total or distinct shaping of the useful part of the instrument*. And it is only from such an experience that a first typical image of the instrumental shape is progressively formed, which makes possible the transition from the work of elaboration to the work of production.

But how could the prehominid ancestor, toward the end of his evolution, begin to shape the edge of both sides of the stone, and thereby prepare himself to rise to the level of humanity of *Homo habilis*? We have seen that the work of elaboration begins with the development of the direct manipulation of the raw material in the transition from the first to the second prehominid phases, when the Australanthropus ancestor began to prepare his instruments in the absence of the object of biological need: from the beginning of hunting expeditions, for example, or when once the game was killed, he went far to look for sharp stones in order to carve it. In both cases, if the *object of biological need* was absent, the *biological situation* nevertheless remained present, and it was this which insured the efficacy of the represented image of the object of biological need, and allowed him to set in motion the preparation and then the elaboration of his instrument. For the need of the instrument could only arise from the diffusion of biological need implied by just such a *situation*. Now it is quite clear that this same biological need situation which conditioned the work of elaboration, at the same time put pressure on the subject, so that he had to remain satisfied with an instrument of a more or less usable shape, not having the leisure time to perfect it. Therefore, as long as the prehominid was restricted to the elaboration of the instrument in the biological need situation, his work could not go beyond the level of a *summary*, or *confused shaping*. The transition to *total*, or *distinct shaping* of the useful part of the instrument is made possible only when the work of elaboration is accomplished *outside of the biological need situation*, in other words, in leisure time, the biological need having already been satisfied.

But what exactly is it that is going to impel the ancestor to work during his leisure time when until now he used it to amuse himself? It is obvious that the simple representation of the absent object of biological need can no longer suffice, since his action presupposes precisely the presence of the need situation. The subject therefore must have a *representation of the biological need situation itself*, in other words, the representation of a whole comprising the object of biological need in its dynamic relation with the prehominid group. This complex image awakens the biological need, and by means of its diffusion, the need to elaborate the instrument, which impels the subject to get to work. Until then, he satisfied himself with summarily shaping a more or less usable thing. But as time is not pressing here, and biological need and thereby the need for the instrument continue to make themselves felt in this *imaginary biological situation*, even when the raw material has already been elaborated in the habitual syncretic shape, the subject finds himself involved in extending his work. In the case of the stone, for example, he starts cutting an edge on the second side, even though the shaping of the first has already given the whole thing a nearly adequate cutting shape.

The total shaping of the edge entirely eliminates the natural shape of that part of the stone and produces for the first time a *distinct* shape, since it is completely worked. In this experience of *creative labor* there is formed a representation of the instrumental shape entirely separate from the natural contingent traits of the raw material, and these are still maintained to a certain extent in the Kafuan cutter where, with the strictly useful traits of the instrumental structure, they formed a syncretic mixture. This first distinct representation becomes more specific by progressive refinement in the practice of collective labor and is formed of a *typical image* of the useful shape of the instrument, as we have seen it realized in the cutting edge of the Olduvian chopper. We can speak for the first time here of an act of *production*, since the worker achieves a result in exact conformity with the aim he assigned himself from the dawn of his consciousness: "At the end of every labour process, a result emerges which had already been conceived by the worker at the beginning, hence already existed ideally. Man not only effects a change of form in the materials of nature; he also realizes [*verwirklicht*] his own purpose in those materials."[29]

The labor of production which has just appeared is not yet tool production since the *tool* must be determined in its *total shape*. The Olduvian chopper, having only its *useful part* determined, still remains in the category of *instrument*.

As opposed to the natural instrument and the prepared instrument, however, whose useful part is determined only on the *functional, non-formal plane*, and unlike the Kafuan instrument or elaborated instrument whose useful part is determined only according to a *syncretic shape*, the Olduvian instrument in its useful part is determined according to a *typical shape*. It, therefore, comes from an incontestable act of production and its author must be classified in the genus *Homo*, since it is the labor of production which defines humanity as such.

Production, however, appears here only in a partial, or, so to speak, embryonic form. As a producer, *Homo habilis* went beyond the *intermediary stage* of ape to man. But as a *producer of instruments, not tools*, he still remains only a "man in the making."[30] He, therefore, still belongs to the *gestation period* of which he will be the final stage, and which will end with "fully-fledged man,"[31] producer of tools. The production of the tool will first appear with the Chellean biface. Then only will man actually detach himself from nature, to emerge in the world of culture, as *Homo faber*. At the Olduvian stage, man in the making, like a fetus in its mothers's womb, developed within the limits of natural existence only; he was not yet a worker. However, he possessed the ability of the first initiator of the productive act, the ability to be a 'jack-of-all-trades' of which we can have a pretty fair idea, considering the abundance and variety of his kinds of instruments. The creator of the Olduvian period thus fully deserves his name of *Homo habilis*.

"Instruments of labour," says Marx, " . . . supply a standard of the degree of development which human labour has attained . . . "[32] In the successive shapes of the instrument and the activities relating to it, we can find a basis for distinguishing, after the anthropoid stage, the principal stages of the *gestation period* of the genus *Homo*.

(1) The *natural instrument*, which sporadically appears among the apes, becomes habitual in usage toward the end of the anthropoid period, which gives rise to the *prepared instrument*. The preparation of the instrument is accomplished in the presence of the object of biological need and is guided by a *generalized sensori-motor image of the instrumental function*. The habit of the work of adaptation entails, on the one hand, the formation of the original indicative sign on the objective plane of the language of real life, and, on the other hand, the transition to erect posture which marks the transition from the *Pongidae* family to the *Hominid* family.

(2) The hand becoming free, *the work of adaptation, with the natural instrument and the prepared instrument, becomes a regular behavior based on the progress of the biological structure*, which implies *the definitive fixation of the indicative sign in the nervous system*. This progress defines the constitution of the genus *Praehomo* as "transition creature" (Engels) between ape and man. The practice of prehominid work immediately entails the cognizance of the indicative sign, which produces the original form of consciousness as *sense certainty*. Within the limits of the natural instrument and the prepared instrument, prehominid work defines the first phase of the intermediary stage or transition stage from ape to man, a phase of which the presently known Australanthropi are late representatives.

(3) The *elaborated instrument* (Kafuan) characterizes the second phase of the transition stage. The work of elaboration presupposes, on the one hand, a *representation of the absent object of biological need* and, on the other hand, a *syncretic representation of the instrumental shape*.

(4) The *produced instrument* (Olduvian) characterizes the final stage of the gestation period (the stage of 'man in the making' or *Homo habilis*). The production of the instrument which occurs during leisure hours implies, on the one hand, the *representation of the absent biological need situation*, on the other hand, a *typical representation* of the instrumental shape, as *shape of the useful part of the instrument*.

(5) Finally the *tool* which appears at the Chellean, marks the *birth*, strictly speaking, of the genus *Homo* in the figure of *Homo faber* (Pithecanthropus). The production of the tool presupposes a *typical representation of its total shape*.

Thus from the characteristic material activities of each stage, we have been able to infer the mental level implied by each. With only the data of prehistory, however, it is impossible to gather information about language which would be indispensable for a study of consciousness during this period. So far, we have established only that the most elementary linguistic sign, the indicative gesture, with its subjectification in the form of sense certainty, and its meaning thus engendered, still reflects the object only in its pure external relation as such, as the pure 'this here' as external reality independent of the subject. And we have accomplished this first step only by recourse to the observation of the child. In order to follow the development of this simple germ of language and consciousness, and to give an effective account of the first forms of representation which so far we have only inferred from the probable evolution of material behavior in the course of *hominization*, we must once more search for concrete data in the psychology of the child.

II. THE BIRTH OF LANGUAGE

Introduction

"For, just as the developmental history of the human embryo in the mother's womb," says Engels, "is only an abbreviated repetition of the history, extending over millions of years, of the bodily evolution of our animal ancestors, beginning from the worm, so the mental development of the human child is only a still more abbreviated repetition of the intellectual development of these same ancestors, at least of the later ones."[33] It is clear that with respect to the intellectual development of the child, it is appropriate to add to our, strictly speaking, animal ancestors, the hominids who have blazed the trail of hominization. Because hominization is still part, in one aspect, of biological evolution, its fundamental stages must be repeated in the maturation of the child, especially so in the maturation of his nervous system, which somehow or other must be reflected in his intellectual development.

The one-year-old period in the child has been called 'the age of the chimpanzee,' for at that stage he can resolve essentially the same problems as the anthropoid.[34] The following stage begins with the indicative sign observable in the child at about 14 months and we have seen that in phylogenesis it marks the transition from the anthropoid ancestor to Australanthropus. We have seen, moreover, that the prehominid ancestor toward the end of his development, already had to avail himself of a certain representation of the absent biological need situation. Now, in the child, it is toward 19 months that a first representation of the absent situation appears, which is demonstrated by the fact that he becomes capable of following a hidden object through *several invisible displacements.*[35] The representation of these successive displacements is possible only within the framework of a complex image which envelops the ensemble of invisible positions, in other words the ensemble of the *invisible situation*, through which the object has been displaced. As a working hypothesis, we thus propose to consider the 14–20 months period in the child as *'the age of the prehominid'*.

On the somatic plane, we can find a confirmation of this hypothesis in the fact that it is exactly when reaching this age that the child becomes capable of walking by himself. At the preceding stage (12–13 months), he was still being supported by the hand of an adult, which does indeed invoke the image of the chimpanzee who walks on his two feet with the help of a stick. At 14 months, bipedal gait is acquired. At the same time, the still clumsy gait of the child, the stiff knees, the legs widespread, somehow remind us that the adaptation to erect posture for the Australanthropus was not fully developed either.

The Developed Indicative Sign

The age of 14 months in the child is characterized by the appearance of a certain number of gestural signs with words used in isolation which have traditionally been called *'word-sentences'*. By this is meant that each of these words would have the meaning of a sentence, which *a priori* seems obvious since we do indeed need a sentence to express a complete meaning.

Actually the problem is not that simple. For if, at this stage, the child could give to his word the meaning of a sentence, it is difficult to see why he would not form the sentence itself. For the meaning of a sentence consists in a judgment, and when we have a judgment, we should be able to join the words in a manner so as to express it. And since the child is not capable of doing this, we are obliged to inquire whether the sentence that is considered equivalent to the word used by itself would not, in fact, express the way the adult understands the child, rather than the way that the child understands himself.

Moreover, we know that each of these words is applied to diverse situations with different meanings. This sometimes disconcerting *polysemy* constitutes one of the major difficulties of the theory of the origin of language. It makes us suspect that the language of the child has its own original semantic structure which is not immediately evident to the adult. And if such a structure could exist, it would be totally undetectable by the mere fact that one would have begun by substituting an adult sentence for the word of the child, which amounts to superimposing an altogether different structure on it.

We believe that in order to avoid the confusion of the adult's point of view with the child's, we must first of all interpret the child's language by itself, in other words by its own 'code'. And since it appears in two stages, one gestural and the other verbal, the gestural being clearer than the verbal, we must begin by directing our attention from the word to the gesture and by analyzing its meaning in terms of its own objective structure.

Let us take a word of this stage that all observers have noted: '*aoua*' or '*avoua*' (*au revoir* – in English: good-bye). Its original meaning obviously comes from the good-bye gesture, consisting in the agitation of the hand which is alternately raised and lowered while being directed toward the person going away. We can easily see that this gesture produces, by tendential projection, an image which reflects precisely the motion of the person going away. This image, thus *signified* by the sign, as a model of the meaning of 'signification', implies in its content three closely knit moments: (1) That of the *object*, produced by the tension of the hand and of the look toward the person going away. Such a tension is but the form itself of the indicative gesture which intends the '*this here*' as external reality independent of the subject. (2) That of the *motion* of this object or person, as motion of the 'this here'. This second moment of the image is projected by the agitation of the outstretched hand. Finally (3), that of the *form* as form of the motion of the object more or less confused with the object itself. This moment consists in the general form of the tendential image produced by the gesture, namely a centrifugal form, or *form of distancing* which results from the projection of the alternating form of the motion of raising and lowering of hand. In short, the meaning of the good-bye gesture consists in an image of the '*this here* (T) in a *motion* (M) in the *form of distancing* (D)', or: TMD,

We can see that the gestural sign of the child, as a linguistic sign, or as a sign relating to the object can be defined as a *developed indicative gesture*. The tension of the hand projects the image of the 'this here', and the motion which is developed with that tension is projected in the image of an external motion which involves the 'this here' in a certain form. We can thus generally say that the meaning of the developed indicative gesture consists in the image of the '*this here* (T) in a *motion* (M) in some form (F), or: TMF.

Such a formula, which we may call a *formula of developed indication*, contains a certain number of possible transformations which will allow us to understand the *polysemy* of the word which reinforces the gesture. According to the characteristic and needs of the situation, the developed indicative gesture will place stress more on one or another part of its structure. As a consequence, one moment will come first, another second, another to a third position in the tendential image thus projected.

The fundamental formula being:

TMF, (1)

we have the variations

TFM, (2)
MFT, (3)
MTF, (4)
FTM, (5)
FMT. (6)

At this stage, the word can take its meaning only directly from the gesture: for, since it is always used by itself, it does not imply any relation to the other words either on the syntagmatic plane, or on the paradigmatic plane. As a result, the transformations of the projected image as meaning of the gesture will immediately give a corresponding multiplicity of meanings to the word. Naturally, the gesture can be reduced to a simple outline or to an internal sketch. But, as we have noted, the outlined motion is as real, as material as the completed motion. And in the total linguistic sign, it is always the whole of the gestural motion, outlined or completed, which, *in the final analysis*, throughout its several mediations establishes the meaning of the word. At the original level at which we place ourselves here, the gesture directly produces the meaning of the word.

It has often been remarked that, at this stage, the word of the child normally contains a duality of meaning, for it is sometimes applied to the object itself in motion as motion of that object. Thus '*avoua*' is applied either to people going away, or to their action of going away.[36] In the list of Gvozdev,[37] we see a child saying '*pici*' (*écris* – write) by showing a pencil. Another time he uses the word to ask his father to draw (he hands him a pencil and a piece of paper). It is commonplace to note that '*apain*' (*pain* – bread) is applied, depending on the situation, either to a piece of bread to be eaten, or to the act of eating it, '*wa-wa*' (water) to the liquid that one drinks, or to the act of drinking, etc.

It is easy to see that this duality of meaning refers to the possibility of the developed indicative gesture emphasizing, in the first place, sometimes the moment of the object, the 'this here', and sometimes the moment of the motion itself. To which it must be added that each of these two cases appears in two possible forms. Thus when it is the moment of the object which comes to the foreground, the gesture, which subtends the word, can stress, in the second place, depending on the situation, either the moment of the motion, or that of the form. And we find once again precisely this double possibility in the polysemy of the child's word. For example, we have just seen that *'avoua'* or *'aoua'* is applied to people going way: 'This here (T) in a motion (M) in the form of distancing (D)', or TMD. But the same word is equally applied to people who have already gone,[38] in other words, who are, so to speak, in the *form of distancing*, as it appears in their motion, past or assumed, of going away, – since an object going away, ends up by disappearing. *'Aoua'* here therefore means: 'this here in the form of distancing (D) as it appears in its (presumed) motion' or: TDM. It should be noticed that in the first case, which is the good-bye gesture, the agitation of the hand is repeated several times, so that the moment of the [going-away] motion itself is clearly marked behind the tension of the hand and arm. In the second case, on the contrary, the subject is happy with just one motion of the hand around the wrist, perpendicular to the direction of the arm stretched toward the object, so that in the tendential image thus projected, it is the form of distancing which is immediately sketched out behind the 'this here', the moment of the motion itself coming only in the third place.

In one of Piaget's observations,[39] we see *'all gone'* [Fr. = *a plus*, for *il n'y en a plus*, literally, 'no more'] being applied by the child to an object that was overturned (without disappearing). Later *'all gone'* merely means that something was at a distance from him (outside of his field of prehension). It is easy to see that in the first case the word means: 'this here (T) in a motion (M) in the form of overturning (O)', or: TMO. In the second case the object outside of his field of prehension is considered, by analogy with the first, as being in that position because of a motion which overturned it. It is thus in the form of overturning, as it appears in the (presumed) motion of overturning it. The meaning of *'all gone'* is therefore: 'this here in the form of overturning (O) as it appears in its (presumed) motion', or TOM.

In an observation of Rosengart-Pouklo, we see a child calling *'koko'* (from *'kuritsa'*, a chicken) any toy having a projection similar to a bird's bill. The word is accompanied by a gesture of the hand mimicking the act of pecking.[40] We can see that the gesture which gives meaning to the word is originally constituted as a developed indication of the pecking chicken: 'this here (T) in a motion (M) in the form of pecking (P)', or: TMP. Afterwards, the same sign is applied to any toy having a bill. But since these toys are, in fact, motionless, the gesture, after first indicating the moment of 'this here' by the tension of the hand and the look, can no longer stress the moment of the motion in the second position. It is thus the form of pecking which occupies the second position. And it appears in the (presumed) motion of the object, since its image is formed by projection from the active gesture of the hand which mimics the motion of pecking. And, in fact, it was indeed that very motion which gave the chicken that characteristic aspect, somehow inscribed in the sharp protruding shape of its beak, which is found in the toys in question. The word '*koko*' here has the following meaning: 'this here in the form of pecking (P) as it appears in its (presumed) motion', or TPM.

In short, the developed indicative sign begins to establish itself *by following the object in its motion*, which takes place in a certain form (F): TMF (1). Then it is generalized to any other object of analogous form, by somehow *reconstituting this form from the motion previously defined*. But since this form is already present in the object, the motion of the hand which develops from its tension, is cut short in such a manner that the emphasis of the gesture is shifted to second place to the moment of the form. It is thus the form of motion which appears at the second level of the projected image; as the form of the object emerging from its virtually present motion, it appears in the third place. The meaning is thus: 'this here in the form (F) as it appears in its (presumed) motion', or TFM (2).

If we now examine the case where the gesture which subtends the word emphasizes primarily the moment of motion (M), we see that it also implies two variations, corresponding to formulas (3): MFT, and (4): MTF. Thus the word '*avoua*' or '*aoua*' can either express the child's desire to be let go when held by the hand or to be taken out of his carriage: it is time to go: '*aoua, aoua!*' (Pichon). But it can also be applied, as noted earlier, simply to the action of going away itself. In either one of these two cases, the tension of the hand is reduced in such a way that it is the developed motion of that tension which plays the major role and consequently is placed in the first position in the formula. In the expression of the desire to go away, the agitation of the hand and arm, in the alternating motion of rising and falling, is intensified to the point where it begins to shake. As a result, insistence in the second place is produced on the moment of the form of distancing, projecting from the alternating form of the shakes of the hand: it is time to go! The moment of the 'this here', which refers back to the child himself taken as an object, only comes in the third place. The meaning is thus: 'The motion (M) in the form of distancing (D) of the this here (T)', or: MDT. On the contrary, when it is simply a matter of ascertaining someone's departure, the agitation of the hand is cut short to the point of being reduced to a simple rotation of the wrist, so that the tension of the arm toward the object is lessened. For the projected image, consequently, it is the moment of the 'this here', which comes in the second place, the moment of the form as the form of distancing coming only in third place. The meaning is thus: 'The motion of the this here in the form of distancing (D)', or MTD.

We can thus say in general that we have a meaning of action in the *imperative mood* when the developed indicative gesture strongly supports a motion of a certain form so as to project the image of a motion (M) of form (F) entailing the 'this here' (T): MFT (3). Inversely the meaning of action appears in the *indicative mood* when the mimicking motion is cut short so as to lessen the tension of the hand immediately. The moment of the 'this here' then comes to the second level of the projected image and the meaning is then defined as 'the motion of the this here in the form (F)': MTF (4).

Of course, what we here call the motion of the object, or motion of the 'this here', must be understood in its most general sense, enveloping every motion of the object, whether it concerns the active motion of the object itself or the motion exerted upon it, or even more generally every motion that concerns it one way or another. Thus *'aoua'* indicates just as much the active motion of someone going away, as the exerted or to be exerted motion upon him so that he will go away (for example, when the child demands to be taken out of his carriage). *'Papo'* (*chapeau* – hat) taken in its sense of action, indicates not only the action of putting on the hat, but also the whole activity of the walk with which the hat is involved.[41] 'Boo-boo' (*bobo* – hurt) designates the object with its active motion of hurting as well as the motion undergone by the child who is hurt. The word *'bébé'* (baby) is used by the child not only with reference to himself and other children, but also with respect to the mirror where he sees the baby. As Pichon expresses it so well: "It is both the fact of seeing himself and the fact of seeing the baby that is contained in the word 'baby'; . . . they are all of the acts concerning the baby."[42] We can see that formulas (3) and (4) of the developed indication can be stated more precisely as follows: 'The movement in the form (F) concerning this here', or MFT (3), and 'the movement concerning this here in the form (F)', or MTF (4).

It is, of course, evident that at this original level, reality is grasped only in its superficial appearances: the relation between the object and the motion is presented only confusedly in the syncretic image projected by the developed indicative sign. Yet, however imperfect such an image may be, it nevertheless constitutes an image *of the real*, in the fundamental structure of all objective reality, as *matter in motion* or *motion of matter*.

The formula of the developed indication still contains two variations, where the moment of the form (F) comes in the first place: FTM (5) and FMT (6). They cannot, obviously, have any actual application here. The disengagement of the form and its positioning in the first place indicate a process which goes beyond syncretism and is only possible at a level superior to the one at which we have arrived. For the first beginnings of language, the developed indicative sign can emphasize only a concrete moment in the first place, that is, the moment of the 'this here' or that of motion. The moment of the form can thus come only in the second or third place.

The Beginning of Language in the Prehominids

As a working hypothesis, we suggested earlier that we consider the period between 14 and 20 months in the child as '*the age of the prehominid*'. We therefore have to examine now whether it would be possible, starting with the living conditions of Australanthropi, to reconstitute the beginnings of language according to a structure more or less comparable to the one just presented.

We must, of course, take into account here a considerable difference: the child comes into the world with the anatomy of *Homo sapiens*, and he has been brought up in a civilized society. Even though his cortex has not completely matured, his first words, at the *prehominid age*, already display the beginnings of articulation, of which Australanthropus was undoubtedly incapable. It is true that the transition to erect posture had to bring about a thickening and rounding of the vocal cords, with a certain softening of the lower jaw, so that Australanthropus probably could emit more varied sounds than the anthropoid.[43] On the other hand, as we have seen, his brain was slightly more developed. However, he kept nearly the same total structure, with the shape regularly rounded, without the protuberances of the endocranium of the Archanthropi and of the Paleonthropi.[44] These protuberances, which indicate the favored directions of the development of the cortex in fossil men, extend in particular around the superior extremity of the sylvian fissure and on the inferior frontal convolution where a little pad called the 'lateral [acoustic] tubercle' is formed. The tubercle, which is near areas 45, 44 (convolution of Broca [i.e., the left inferior frontal convolution, usually more highly developed than the right, and discovered by Broca in 1861 as the center for language]), which make possible the motor mechanism of articulate speech in present-day man, essentially involves the inhibition of sounds immediately after their emission which allows one to differentiate them by passing distinctly from one to the other.[45] Now, the lateral tubercle does not yet exist in the Australanthropus. We can infer from this that Australanthropus did not yet exhibit this phenomenon of vocal inhibition, so that each sound he emitted was more or less prolonged in a *diffuse* form.

Articulation was thus impossible, and the vocal emissions of Australanthropus must have been comparable to those of the anthropoid, although undoubtedly more varied.

In the ape, it is still only by means of signals that the emotional aspect of action is expressed. We have shown in the preceding section that sound emission took on an objective, or linguistic meaning for the first time by being linked to the original indicative gesture. It could be that the '*Ah*!' which in the child indicates the '*this here*' comes from the '*ak*' sound that the ape emits when confronted with a disquieting situation, and which the anthropoid ancestor, toward the end of his evolution, repeated when, using his hand, he began to indicate the object of collective labor. The constitution of the first linguistic sign by the union of the emotional sound with the indicative gesture marks the transition from the *pongid* family to that of the *hominid*. With the appearance of Australanthropus, the process is continued with the very development of the indicative gesture: in the structural study presented earlier, the examples given such as '*aoua*', the sign of distancing, 'all gone', the sign of overturning, and '*koko*', the sign of pecking, can be easily transposed to the conditions of phylogenesis.

We have seen that, most likely from the very beginning of prehominid development, in the cognizance of the indicative sign, the original form of the circular arc gesture was transmuted into the straight line form. Yet if, by virtue of the excitation of collective work, the straight line indicative gesture is prolonged for an instant, *the prehominid necessarily follows the object in its motion*: for example, the game that is fleeing or falls down, or the bone fragment or piece of wood which pierces the animal like a beak or a dagger. The gestural sign developed in this way is reinforced each time by a *diffuse sound*, of emotional origin, but which is now related to the tendential image projected by the gesture, and in this way obtains value as a *word* with an objective meaning: '*this here* in a motion in the form of distancing, overturning, piercing', etc. The developed indicative sign has thus been constituted in its fundamental semantic formula, TMF (1): 'this here in a motion in the form (F)'. It is evident that the communication of such a meaning content allows a coordination of collective labor by far superior to the simple concentration of the forces of the group on the object indicated as the 'this here!' As a result, the new sign is repeated in order to indicate the moving object in diverse forms. On the other hand, in the process of social practice, each developed indicative gesture can vary slightly in different ways depending on the character and needs of the situation. Each sign, as unity of gesture and word, thus acquires a polyvalent meaning which encompasses the four semantic formulas described earlier: the object in a motion of a certain form, TMF (1) – the object in a certain form as it appears in its (presumed) motion, TFM (2) – the desired motion concerning the object, MFT (3) – finally the motion ascertained in the object itself MTF (4). These variations are not yet clearly distinct in themselves, since, stemming directly from the various nuances of the gesture, they can be distinguished only in relation to the concrete situations on which the gesture itself depends. In the meaning of the total sign, they are thus more or less confused in a *syncretic* ensemble which cannot be differentiated except according to the situation. Thus, if we consider more particularly the verbal moment, we can say that the word appears at this level as *diffuse* on the plane of the signifying and *syncretic* on the plane of the signified. In short, to the extent that we can represent it to ourself, language appears in the prehominid in a *non-articulated form*, made up of *developed indicative gestures* reinforced by *diffuse-syncretic words*.

Until now, the transition from the simple indication to the developed indication has been accomplished only on the objective plane of the *language of real life*, by a *modeling of the signifying act*, in other words, from the motion of the hand stretched toward the object under the material conditions of collective labor, which leads the subject *to follow the object in its motion*. As a result of this, the signified, or the tendential image thus projected, reflects the object no longer simply as a 'this here' as such, but also in a motion of a certain form, as it appears in the material activity and the material relations of the prehominids. On the plane of consciousness, however, we still have only the original intentionality of sense certainty as immediate intention of the 'this here'. The moments of the motion and of the form still appear in the signified image only as *tendential moments* which are objectively communicated by the subjects to one another in social practice, but still do not imply any consciousness: consequently they appear only by virtue of the situation, so that the subject cannot avail himself of them in himself.

The cognizance of the developed indicative sign is produced only when the *subject addresses it to himself*. This can be seen in Konnikova's observation of a child *at the prehominid age*: The word '*taka*' originated in the child the moment that his father, using a stick, reached a toy that had rolled under the couch. Several days later, observing someone sweeping under the bed, the child once again says '*taka*' – and the broom received that name. After that, while playing with toys on the floor, the child threw a pencil under the couch and smilingly said '*taka*'. Then the world '*taka*' was applied to a knife, and finally to a long ribbon.[46]

It is clear that here the word gets its meaning from a sudden hand gesture of reaching which mimicked the motion of the stick, then of the broom which the child saw sliding under the couch or under the bed. It is a question, so to speak, of a 'lengthening in depth'. '*Taka*' thus began by meaning: 'this here in a motion in the form of lengthening in depth (Ld)', or TMLd. Then he indicated the broom at rest as a 'this here in the form of lengthening in depth (Ld) as it appears in its (presumed) motion': TLdM. Thus, when the child later threw the pencil under the couch, by using the same word '*taka*', the action of his hand in reality fulfilled a double function. On the one hand, it repeated the same developed indicative gesture, but also emphasized the moment of motion in the form in question to be executed on the object: MLdT. In other words the child *indicated to himself* the sliding of the pencil under the couch as he had seen his father do with the stick, and the word reinforced the gesture. On the other hand, by the same action of the hand, he executed that order which he addressed himself. In this way the material motion of the sign realized its own meaning, which defines the very concept of *play*, for play is a signifying act which realizes its own significance. This was noticeable in the *smile* of the child. Through this active operation of *cognizance*, the sign addressed to himself becomes for him an available acquisition, which leads him to extend its field of application by continuing to focus only upon the general form of lengthening. And it is in this way that he designated the knife, then the ribbon as the 'this here in the form of lengthening (L) as it appears in its (presumed) motion': TLM.

If we go back to phylogenesis, we see that this return of the sign to oneself, which arose in the child in the form of play, had, in fact, its original conditions in the practice of collective labor. The prehominid workers sent one another the same developed indication of the object according to a motion of form (F). But if one of them lagged behind, the sign which he automatically repeated necessarily returned to himself, since in his position as lagger, he in reality did not have to call the others to that work, so that it is in fact *for himself* that he repeated the call which came to him from the others. Having identified himself with them in that call to himself, *he sees himself in them as calling himself to action in that form* (*F*) *to the object*. Afterwards, once the various individuals have occasionally experienced this relation to the self, as return to the self of a relation to another, the new structure becomes generalized in a *collective cognizance*, where each one, while returning the sign to the others, at the same time addresses it to himself. Consequently, the sign is, so to speak, carried by social reciprocity returning upon itself in the group, so that it becomes available to the group itself, independently of any immediate task. Finally, in due time, the enduring image of the group in each individual suffices to make him recall this sign, so that it is *reactivated* as soon as an object elicits, in the dynamic field of perception, a tendential figure which re-enters one of the possible variations of the meaning already acquired. Consequently, the subject constantly avails himself henceforth of that sign in the enduring image he keeps of his group, and in which he sees himself *in himself*, which amounts to saying that he avails himself of it in his *consciousness.*[47]

The First Signs of Representation

The cognizance of the developed indication obviously makes for considerable progress in the organization of labor. The appearance of that sign on the objective plane of the language of real life had already furnished the pre-hominid group with a means of coordinating its efforts no longer simply by concentrating them on the same object indicated as the '*this here*', but also by orienting them, directly or indirectly, to the action adapted to the needs of the situation. The orientation is given directly when the sign is exchanged in the imperative mood according to formula (3): MFT, and indirectly in the three other cases – formulas (1), (2) and (4), – which are in the indicative mood. Once the sign becomes conscious, it enables the subject to explicate *for himself* the content of his perceptive field. This sign thereby awakens *individual initiative* at the same time that it appears as a means available to the individual to mobilize and orient collective labor.

Such progress, however, does not yet bring any essential transformation in the very structure of the work of adaptation, as it was constituted in ordinary behavior with the birth of Australanthropus. Until then, the total instrumental activity remained centered on the perceived object, so that its development proceeded only within the limits of present perception. We are still dealing here with only *quantitative* progress. The *qualitative leap* is realized only with the transcendence of the present perception through the beginning of *representation*. Indeed, we have seen that the transition from the preparation of the instrument, by simple direct manipulation of the raw material, to its elaboration by means of a second instrument was possible only when Australanthropus began to prepare his instruments at a distance from the object of biological need, which implies the representation of that object. On the other hand, the preparation of the instrument under these conditions also required a certain representation of the instrumental form. The development of language and consciousness in the first prehominid phase must have therefore ended in this double representation, thanks to which, fundamental progress in the structure of the work of adaptation was realized, and this marked the transition to the second phase.

A. *The beginnings of representation in the child.* We can observe the origin of representation in the child the moment he shows himself capable of finding behind a screen an object which was placed there when he was not looking. This behavior was described for the first time by Piaget who observed it in two of his children, one (Jacqueline) at the age of 18 months, and the other (Lucienne) at the age of 13 months. The observer puts an object in his hand, and closes it before the child's eyes. He then puts his hand behind a screen where he places the object. He brings his hand out closed and empty. The child searches for the object in his hand, and not finding it there, hesitates, gropes, and finally searches behind the screen.[48]

Such a performance assumes in the subject something more than a simple enduring sensori-motor image of the object associated with the perception of the screen, as is the case when the object is hidden while in view of the child. For example, when the child searches for it in the observer's hand, which he has just seen being closed, this hand, which functions as a screen, is at the same time perceived as an *indicator* which evokes this object, so that, although invisible, it nevertheless remains somehow *still present* in the perceptive field. In other words, the sensori-motor image of the object is maintained in the form of an *enduring image* which, through the mediation of the indicator, continues more or less to be part of present perception. We can see that the search for the object behind the screen, in the case where it has been put there with the child watching, does not go beyond the limits of the field of presence, in the broad sense, of the sensori-motor perception, and, consequently, does not assume any '*re-presentation*' as *imaginary presentation* of an absent object. This searching, which amounts in fact to detour behavior in mammals, appears in the child about the age of 9 months, or at the fourth stage of sensori-motor intelligence in Piaget's classification.

The situation is altogether different when the object, unseen by the subject, is hidden, as in the observations mentioned earlier. In this case there is no indicator in the perceptive field to evoke the presence of the object behind the screen, so that the child can no longer guide himself by a simple enduring image. In order to find the object, under these conditions, the child must 're-present' it to himself, in other words, he 'presents it again' to himself outside of the field of presence of present perception, which is possible only through an active image which marks the beginning of *thought*.

It is true that Piaget attributes the behavior of his two subjects not to the intervention of representation, but rather to practical learning, which can be seen in the groping behavior which preceded success. It would be precisely for this reason, at this level, that the child masters *only one invisible displacement* of the object. If, after the success of the experiment in question, the observer puts his closed hand behind a second screen in order to place the object there, the child again searches for it behind the first screen and not behind the second.[49] It would seem then that the preceding performance refers only to a sensori-motor scheme acquired by empirical learning. According to Piaget, the existence of representation can be considered as established only when the child is capable of mastering *several invisible displacements*, the observer having successively put his closed hand behind several screens in order to place the object only behind the last one. This experiment was finally successful for Jacqueline at the age of 19 months and for Lucienne at the age of 15 months,[50] which indicates a clearly higher level of development.

Howevers, on the basis of the results of systematic observations of 90 children from 3 to 20 months, Gouin-Décarie has shown that owing to the statistical distribution of the performances among the different levels, it would be more logical to admit the existence of a beginning, obviously still unstable, of representation *from the mastery of a single invisible displacement* of the object. The mastery of several invisible displacements, which appear at the following level, would indicate, not the origin, strictly speaking, but rather the stabilization and the reinforcement of the representation.[51] This objection becomes all the more weighty in view of the fact that it has been raised right inside of the Piaget school, Gouin-Décarie having taken Piaget's system as the basis for her research.

I have repeated the observation of this behavior with a little boy of 16 months. We were sitting face to face on a mat. Having shown him a little celluloid ball, I passed it behind my back, and hid it under one of my coat-tails that was resting on the mat. The child immediately got up and headed toward me. He leaned over to look at the bottom of my back, pointing his finger at it, and remained motionless for about ten seconds. Then he bent down, lifted my coat-tail and took the ball.

We can see that here the phenomenon of groping on which Piaget insisted, is replaced by an indicative gesture pointing to the object, it would seem, *through the screen*. But how is this possible? Until now we have seen that the indicative motion concerns the '*this here*' as objective reality given to sense intuition. Apparently, it should not then go beyond the field of present perception. Yet, in the observation just mentioned, not only was the object invisible under my coat-tail, but what is more, there was not any sign that could evoke its presence in that place, since I hid it there without the child knowing about it. How then could he *indicate* it with his finger?

A few minutes before this observation, within view of the same child, I had placed the little ball under an overturned box, the open end facing down on the mat. The child pointed his finger toward the middle of the box while looking at me, as if he expected me to take the ball from under the box and give it to him. The indicative gesture here is based on an enduring image, mediated by the perception of the box, which functions at the same time as sign and as screen. In other words, the box evoked the presence of the ball underneath, so that, although invisible, it nevertheless remained somehow *still present* in the present field of perception: in this sense it was 'still perceived' under the box, so that the subject could indicate it by pointing with his finger which aimed at the ball *through the box*, as a '*this here still present*'.

Thus when I passed the ball behind my back, the child kept the enduring image of it mediated by the perception of my own body behind which he saw it disappearing, and also by the perception of the mat, since my hand had slightly touched it while shifting the ball. In the child's perceptive field, in the broad sense, the enduring presence of the ball evoked by that double sign was thus more or less localized at the bottom of my back. And this is what motivated the motion of the child when he headed toward me in order to look in that direction. Thus when he pointed toward it with his finger, that indicative sign was concerned with the same enduring presence: in other words *the child indicated to himself the ball that he did not see, but whose enduring image he retained as located at the bottom of my back*. The gesture of the pointed finger thus did not, at that moment, aim at the object through the screen formed by my coat-tail. After ten seconds, in that position, however, the child leaned over in order to lift the screen. The gesture of the pointed finger thus took on a new meaning: he now aims at the object *through the screen*, as being underneath it. The situation apparently offers some analogy to the preceding observation where the child indicated the ball under the box, with the altogether fundamental difference that the box functioned as a sign which evoked the ball as 'still present' in the perceptive field. My coat-tail, on the contrary, could not evoke anything in the eyes of the child, since I had hidden the ball beneath it without the child knowing it. In this position the ball could be aimed at only as being outside the field of presence of present perception, or as 'absent'. We can see that during these ten seconds, which we can consider as a period of *reflection*, an essential transformation occurred. At the beginning, the gesture of the pointed finger was aimed at the ball according to its enduring image at the bottom of my back, so that the child indicated it as a '*this here still present*', even though in fact it was invisible in the perceptive field. Now he indicates it under my coat-tail as a '*this here absent*' but '*represented*' or 'presented again' outside of present perception. The direction of the pointed finger, in other words, the external form of the signifying, does not seem to have changed, but the thing signified has become altogether different. How could one instant of '*reflection*' achieve such a result?

In our preceding section we saw that the linguistic sign and above all the indicative sign are constituted in the dialectic of social reciprocity which they continue to carry in their own immanent structure. The new meaning which has just appeared, thus necessarily refers back to a progressive development of the reciprocal structure of the sign itself. It would be difficult, however, to study it directly in the child since he only gradually brings into play a heritage acquired in the social past of the *hominid family*. We must, therefore, return to phylogenesis.

B. *The origins of the sign of representation in prehominid development*. As we remarked earlier, the search for the missing object appears in its most simple form in mammals, in detour behavior, which already implies the extension of the perceptive image of the object in an enduring image. With the appearance of Australanthropus, the development of the indicative gesture undoubtedly made possible, in the course of the first prehominid phase, the indication of the missing object by its enduring image at the very place where it disappeared, or better still where the subject is accustomed to see it disappear or appear as we can observe in the child who, toward the age of 15 months, points his finger toward the door when asked 'where is mommy?' Thus, if in the course of a hunt the game flees behind a mountain, the prehominid hunters could indicate it to themselves by pointing the hand toward the curved side of the mountain where it has disappeared, but where it *still remains present* by its enduring image.

Afterward, however, this first indicative form of the 'this here still present' is modified according to the real situation. Thus a few hunters, having gone ahead, may arrive at the turning point while the majority are still in front of the mountain. The leaders then perceive the game on the other side of the mountain and indicate it to the others by a gesture of the hand pointing ahead, behind the mountain. The others respond by the same sign, now guided by the tendential motion of the hand of the leaders, by following with their eyes the straight line which extends the gesture of the leaders toward the game. As a result of this, the indicative sign by which the main body of the troop responds to the leaders, takes a more or less *perpendicular* direction to the sign of the leaders. In other words, the hunters who remained behind, in front of the mountain now indicate the game, no longer at the curve of the mountain side where they saw it disappear, but ahead, *on the other side of* [behind] *the mountain*, by a gesture of the hand which aims at it *through that screen*. In other words, because of the new situation, the reciprocity of the sign is modified in such a manner that the image of the '*this here*' projected by the gesture of the outstretched hand, is shifted, carrying along with it the enduring sensori-motor image of the game, which was first at the curve of the mountain side and is now transferred ahead to the other side of [behind] the mountain.

By a sort of modeling of the material conditions of collective labor, a new form of the sign is constituted, which we can rediscover in the child's gesture described earlier, when he indicated the ball hidden under the box by pointing his finger toward the middle of it and not toward the edge of the box under which he saw the ball disappear. We can see that, in the child, this first indication of the object through the screen already implied a certain progress in the internalized reciprocal structure, inherited from his origins, which became immanent in the sign itself, all of which involves a transfer of the enduring sensori-motor image of the object under the box from the edge toward the middle. We are not yet dealing here, however, with the birth of a new structure, but rather with a simple variation of the same structure. In effect, the enduring image of the ball, which the subject saw slipping under the edge of the box, was already more or less inside of that edge, under the box, so that its transfer under the middle of the box, does not bring any essential change to the relation of all the positions. The case is the same for the enduring image of the game which was located more or less around the curve of the mountain side and consequently behind the mountain itself, so that its transfer ahead does not essentially modify its position in relation to the main body of the hunters, for whom the screen formed by the mountain continues to function as a sign evoking the actual presence of the game behind it. As a result, as in the case of the ball under the box, the object aimed at through the screen is always indicated as a 'this here still present' in the field of present perception. In other words, we always remain within the limits of the original structure of the indicative sign, insofar as it aims at the *this here* as objective reality given – or 'still given' – to sense intuition. There is thus as yet no occasion to speak of *representation*.

Meanwhile the game, which continues to flee behind the mountain, hard-pressed by the leaders of the Australanthropi hunters, is suddenly faced with a boulder. It immediately goes around it, and once more escapes out of sight of its pursuers. They arrive in their turn, go around the same bend and see the animal once again behind it. At the instant immediately preceding, the moment when the animal disappeared behind the boulder, the main body of the hunters had just reached the first bend. They look behind the mountain toward the place where they had pointed the game out to themselves by guiding themselves by the indicative sign given to them by the leaders who were then at the same position where they are now. And as they do not see anything, they stretch out their hands aiming at the animal, who, though invisible, is *still present* in its enduring image. They remain in this position looking for the animal but only see the boulder, which the animal had gone around without their knowing. At this point, the leaders who have just arrived at a second bend to the side of the boulder, indicate to them, by pointing their hands, that the game is behind the boulder. Immediately thereupon, the gesture of the main body of the hunters takes on a new meaning. For it now responds to the new sign of the leaders, and guided by it, indicates the game *behind the boulder*. Consequently, the enduring image of the animal is transposed to a position behind that second screen.

We apparently have here a situation nearly comparable to the situation described earlier, at the time when the majority of the prehominid group responded to the gesture of the leaders at the first bend and thus indicated the game *behind the mountain*. However, an altogether fundamental difference has been introduced. In the preceding case the mountain continued to function as a sign, so that the gesture which aimed at the game through it, always indicated it as a 'this here still present' in the perceptive field. And this meaning of the 'still present' presence of the object is maintained, when the majority of hunters, having arrived at the first bend and not seeing the animal behind the mountain, tried to gesture to the animal by means of its enduring image. But now that the same gesture, responding to the sign of the leaders at the second bend, indicates the game behind the boulder, the meaning becomes altogether different, since the boulder does not function as a sign for the majority of the hunters. The indication of the object behind that second screen, without any sign evoking its actual presence, can be intended only as a '*this here absent*', but '*re-presented*', or '*presented again*' outside of the perceptive field in a reciprocal relation with the gesture of the leaders who directly indicate it in that place.

Thus the modeling of reciprocal signification in the material conditions of collective labor ends this time no longer in a simple variation inside the same structure, but in an altogether new structure, the indicative sign now functioning as the *sign of representation* or *representative indication*. As opposed to this, the indication within the limits of the field of presence of present perception may be called the *sign of presentation* or *presentative indication*.

The new meaning which has just arisen on the objective plane of the *language of real life* has the peculiarity that it immediately implies, by its objective form, the moment of subjectivity. We have seen in the dialectic of the original indicative sign that cognizance begins in a sporadic form in a situation of lagging behind, where the subject does not actually have to call the others to the object, since he himself is lagging behind, so that the sign, which he automatically sends back to them, returns to himself and is confused with theirs, from which it results that it is in fact to himself that the subject addresses that call from the others with whom he identifies. In the representative indication of the absent object, as it just appeared in the Australanthropi hunters, we immediately find ourselves in a similar situation since the main body of the group which issues this sign is lagging behind the leaders, and does not, consequently, have to call them to pursue the game. It is thus in fact to *themselves* that the main body of the hunters address this indication, in other words they call one another and each one calls himself by identifying himself with the call that comes from the leaders, among whom they see themselves. Consequently, the new sign has immediately its own image *in itself*, so that the main body of the hunters of the group *recognize themselves* in the call of the leaders, from whom they call themselves to the pursuit of the game behind the boulder. Now it is the very movement of such a recognition which constitutes the form of its *lived experience*, as *consciousness* for the signifying material act. We can thus say that the representative indicative sign appears from the beginning as *conscious*.

Obviously we are dealing here with a merely sporadic consciousness, tied to rather particular circumstances in the course of collective labor. The representative lived experience is formed as a permanent structure only when the new form of reciprocity is internalized in the enduring image of the group, so as to differentiate within it a particular moment, the *moment of the leaders*, whose role is constantly to outline the indicative sign from the edge of each object by aiming at a point behind it. It follows from this that the subject, who always keeps the social image within himself in the identity of his own lived experience, constantly sees himself *indicating something behind the object he is looking at.* As a result, perception from now on is fitted into a new structure where it is broadened by the moment of *representation* which constantly projects *beyond the limits of the field of presence of present perception* the more or less indeterminate image of a distant space, but which must be made progressively more precise in terms of the real situation and activity. Such a structure is stabilized in the hereditary form of the prehominid ancestor's nerve synapses toward the end of the first stage of his development, and it is their reactivation in ontogenesis that we have witnessed in the observation of the child who pointed the finger, while searching for the ball at the bottom of my back, and then, after a few seconds in that position, leaned over in order to take it from under my coat-tail. In this short time lapse, which seemed to us a moment of *reflection*, the transition from the presentative indicative sign, in its broad sense, as indication of the 'this here still present', to the first form of the sign of representation as the indication of the 'this here absent' was made. 'Reflection' (which we can define here as the continuation, for a while, of the experienced return of the sign to oneself in the internalized movement of social reciprocity) has led to the differentiation in this 'immanent environment' of 'a moment of the leaders', thanks to whom the gesture of the pointed finger takes on the meaning of an indication of the absent object behind a screen, since it now responds internally to the outlined image of another indication which, from the hem of my coat-tail, aims at the ball underneath. The appearance of this new structure of the sign undoubtedly had been prepared for by the social experience of the child, who on various occasions was able to see people point out to him an object behind a screen, put there without his knowledge. However, the very position of this appearance in child development, marks a determinate stage and thereby refers back to a corresponding stage in phylogenesis. In fact the subject of which we speak was 16 months old. Earlier we recalled Piaget's observations of the mastering of the invisible displacement of a removed object by two of his children, one at 18 and the other at 13 months. Gouin-Décarie has noted the same behavior in a subject of 20 months (see Notes 35 and 51). Thus if we take the average of these ages we can place the level of this behavior at about 16–17 months, or nearly in the middle of the *prehominid age* (14–20 months), which does indeed correspond to the time when the transition from the first to the second phase of prehominid development in prehistory begins.

We have shown in the first part of our study how the *freeing of the hand* by the acquisition of erect posture was the *'decisive step'* which opened the way for *hominization*, thereby marking the transition from the Pongid family to that of the hominids. We can now consider the appearance of the sign of representation at about the middle of the prehominid stage, or transition stage between ape and man, as a true *freeing of the brain*, whose superior functions, language and consciousness, from now on can transcend the narrow limits of the present situation, where animal psychism is imprisoned, which the presentative indicative sign had not yet transcended.

The freeing of the brain was the *second decisive step* which opened the way for the indefinitely progressive formation of an image of the world in its universality, as infinity of matter in motion, and thereby allowed man to consolidate more and more each day his domination of nature and himself. This same going beyond the present given, however, also opened up the possibility, from a certain level of development of language and consciousness, of completely escaping beyond the real in order to be enclosed in purely symbolic constructions which deny the reality of human life and overturn its sense of truth. Idealism justifies the principle of these constructions by interpreting the concept of 'going beyond' as a pure negation of objective reality, which, in the final analysis, can end only in a mystical 'transcendence'. The concrete analysis of the sign of representation allows us to specify the authentic meaning of *going beyond* by making explicit its actually real content, that is, its *social* content. The going beyond the field of presence of present perception is already itself objectively *given* in the social activity of labor, where the group broadens its field of action by dividing itself into different sub-groups, at the same time that, through linguistic communication which assures the coordination of tasks, it unites all of its members in the form of one and the same *'collective worker'* which is thus found *'present everywhere'* in an increasingly vast area of natural reality. " ... a body of men working together," says Marx "have eyes and hands both in front and behind, and can be said to be to a certain extent omnipresent."[52] And it is, of course, this *real 'omnipresence'* of the collective worker, true 'Briareus'[53] coupled with an Argus, which produces the *ideal omnipresence* of consciousness, when the subject, starting with the enduring image of the group which he possesses within himself in the form of his lived experience, repeats to himself the sign of representation in increasingly complex structures which reflect the real progress of social labor in its constantly open perspectives, and thus indefinitely *goes beyond* himself *in himself*.

We have described the beginning of representation in its most elementary form, as the simple indication of the 'this here absent'. And we have seen that in these conditions it implied the present presence of an enduring image, transposed beyond the perceptive field by the gesture of the hand aiming at the object through the screen. Thus in the child, who pointed to the invisible ball as 'still present' behind my back, the intervention of the new structure which gives this gesture the meaning of an indication of the object under my coat-tail has, thereby, caused the transposition of the enduring image of the ball under that screen. It is clear that such an operation is possible only for an object which has only just disappeared, and whose enduring presence is still maintained rather clearly in present perception. As the missing object plunges into the past, its enduring image tends to be erased from the perceptive field, so that the sign of representation can function effectively only on the condition of *reactivating* it.

Now, this was undoubtedly the problem that came up in phylogenesis at the end of the first phase of prehominid development, when Australanthropus began to prepare his instruments in the absence of the object of biological need. If the representation of the object, which first occurs at that epoch, had been limited to its simple indication as a 'this here absent', the margin of freedom thus assured for the search for material would not have been very great, since the sign, as we have just seen, implies the presence in the perceptive field, of an enduring image of the object still relatively precise. Thus as the Australanthropi hunters tended to increase the distance from their biological object in order to find the material necessary for the preparation of instruments, there had to be a more detailed indication associated with the simple gesture of the hand aiming at the biological object as a 'this here absent', so as to refresh its enduring image already more or less faded. Now, at the level we are considering, the subject has at his command the developed indicative sign which in formula (2) TFM, projects the image of the 'this here in the form (F) as it appears in its (presumed) motion'. Until now this sign functioned only within the limits of present perception. We may believe that owing to the exigencies of the new situation, it is now integrated into the structure of representation so that from now on it has two components, one indicating the 'this here absent' and the other representing it in the aspect under which it is usually presented in the practice of collective work, in other words, as a 'this here in the form (F) as it appears in its (presumed) motion.

C. *The composite indicative sign*. We can observe in the child at the first stage of *the prehominid age*, the association of the simple with the developed indicative sign within the framework of present perception. Thus in one of Piaget's observations[54] we see a little girl of 13 months 29 days saying "*bow-wow*" as she pointed at the geometrical pattern on a rug (a horizontal line crossed by three vertical lines). The same word had been used previously to designate dogs. Such a polysemy clearly shows that it is not simply a question of onomatopoeia. The word '*bow-wow*' gets its original meaning from the developed indicative gesture pointing to the walking dog and seeing it as a 'this here in a motion in the form of the projection of the paws (P)', or TMP. And it is precisely this same form that the child rediscovers in the shape of the object in the outline of the horizontal line crossed by three vertical lines at which she pointed her finger while repeating "*bow-wow*." Thus we are dealing here with a double sign, consisting of the pointing finger gesture indicating the 'this here' (T), and the syncretic word '*bow-wow*' which implies an outlined gesture indicating the 'this here (T) in the form of the projecting of the paws (P) as it appears in its (presumed) motion (M)', or TPM. The total meaning is thus T. TPM, which is stated as follows: '*this here as a this here in the form of the projection of the paws as it appears in its (presumed) motion*'. We can thus call this sign the *composite indicative sign*.

We can find this same structure again in another observation, by the same author, of his little 16-month old boy who said "*mummy*" as he pointed to everything that he wanted, even when he was referring to his father or to some other person.[55] We can see that here '*mummy*' is a syncretic word which gets its meaning from a developed indicative gesture aiming at a movement in the form of bringing close the 'this here', that is, the desired object already indicated by the simple gesture of the pointed finger. I have observed a little girl of 14 months standing on the door step and looking into the street. The noise made by a passing car having scared her, she turned around crying to her baby-sitter and stretched her hand to her saying "*mummy*", the hand in a pronated position making a rather slow alternating gesture of raising and lowering at the wrist. It is almost the same gesture as the good-bye gesture, but executed slowly, which projects the image of a movement of drawing the indicated object or the person nearer, as opposed to the image of distancing projecting by the rapid agitation of the hand waving good-bye.

We see that the word *'mummy'*, which accentuates the gesture, is a call for help, whose meaning is defined according to formula (3) of the developed indication (imperative mood): 'the motion in the form of drawing (D) closer to this here' or MDT, the 'this here' being here the baby-sitter. We may believe that the word *'mummy'* uttered by Piaget's subject implies, in an outlined form, a gesture of the same kind. Its association with the simple gesture of the finger pointed toward the desired object constituted a composite indicative sign, whose meaning is; T. MDT, '*this here in the motion in the form of drawing closer to this here*', or more briefly: '*this here in the motion in the form of its drawing closer*'.

We see that the developed indicative sign which constitutes the second element of the composite indicative sign, can be presented according to any of its possible formulas. Thus if we formalize the composite indication, we can use for its second member the fundamental formula of the developed indication – TMF (1) – it being understood that it can be replaced by one of its variations. As the *general formula of composite indication* we have:

$$\mathrm{T.TMF}, \tag{7}$$

which is stated: '*this here as a this here in a motion in the form* (F)'.

D. *The general formula of the representation of the absent object*. We have seen that at the end of the first phase of prehominid development in phylogenesis, as the Australanthropus ancestor tended to prepare his instruments farther away from the object of biological need, its representation, which first appeared in the form of a simple indication of the 'this here absent', undoubtedly became more precise by uniting a developed indicative sign with it. We may believe now that this association appeared as a composite indicative sign, such as we have just described, integrated into the structure of representation: in other words a gesture of the hand stretching into the distance, indicating the invisible object in the swamps, with the syncretic word ordinarily used to designate it in the habitual motion concerning it. The meaning of this complex sign can be given by adding to the general formula of the composite indication (7) a supplementary symbol, for example, a line drawn above the letters, in order to indicate that it is not a question of a present object, but of an absent object *represented* outside of the perceptive field:

$$\overline{\mathrm{T.TMF}}, \tag{8}$$

which is stated: '*this here* (*absent*) *represented as a this here in a motion in the form (F)*'. As we remarked earlier, the second member of this expression which enters into it as the fundamental formula of the developed indication, can always be replaced by one of its variations. In the representation of the absent object of biological need, it may appear motionless – for example when the game has already been killed. The developed indicative sign which enters into this representation, will in this case follow formula (2), TFM, and the total meaning will be: $\overline{\mathrm{T.TFM}}$, '*this here (absent) represented as a this here in the form (F) as it appears in its (presumed) motion*'.

As the simple indication of the 'this here absent' very much depends on the enduring image of the missing object, and as nothing is learned about it, except that it really exists outside of the perceptive field, we can consider it as an altogether embryonic form: we thus propose to call it the *provisional sign of the representation of the absent object*. This representation becomes firm only when it becomes more precise by the addition of a syncretic word which adds the indication of a motion and a form concerning this object. It is only then that we have, strictly speaking, the *sign of the representation of the absent object*, so that formula (8) can be considered as the *general formula of the representation of an absent object*.

Now, this formula can help us understand the syncretic representation of the instrumental form, which, in the first part of our study, was seen to be necessary for the preparation of the instrument far from the object of biological need and for the transition to the work of elaboration. In effect, if we examine the symbol introduced in this formula, that is, the line drawn above the letters designating the representative structure wherein the first two components of the total meaning are integrated, we see that it could very well affect only one of them. Thus we would have two new possible formulas: $\mathrm{T.}\overline{\mathrm{TMF}}$ and $\overline{\mathrm{T}}\mathrm{.TMF}$. The first, considered in its variation $\overline{\mathrm{T.TMF}}$ is altogether in accordance with the syncretic representation of the instrumental form. Whereas the second can take the form $\mathrm{T.}\overline{\mathrm{MTF}}$, which enables it to be applied to the phenomenon of deferred imitation which we also encounter in the child at the level considered here.

E. *The sign of syncretic representation of the instrumental form*. We have seen that the preparation of instruments which occurs in the anthropoid ancestor toward the end of his development was only guided by a generalized sensori-motor image of the instrumental function, in other words, an image of the characteristic motion of the instrument to be prepared, immediately applied to the perceptive image of the present object of biological need. With the transition to the prehominid level, the indicative sign had to intervene in this process. The Australanthropi probably indicated to one another, and each one to himself, the raw material to be used by stretching their hands toward it. Such a gesture, more or less repeated, was enough to bring agreement as to the choice of material, since the subjects already have at their disposal a functional image of the instrument into which the material has to be transformed. It is true that in the very course of that preparation, the subjects still had to encourage one another with a developed indicative sign, probably the same as the one they already used to designate the instrument with the usual movement related to it. If, for example, it is a question of a cutting instrument, the fundamental meaning being 'this here in a motion in the form of cutting (C)', or TMC, we will obtain in the present case, in the imperative mood, according to formula (3): MCT, 'the motion in the form of cutting concerning this here'. There occurs here a transposition, altogether normal at the syncretic level, from the useful motion characteristic of the instrument, to the motion characteristic of its preparation for the manipulation of material, in other words, a confusion between the motion of manipulation which must allow the stone to cut, and the very motion of cutting with that stone. Consequently, the image projected here by the sign-as-meaning is confused in practice in its content with the functional image which had already guided the preparation of the instrument, namely the generalized sensori-motor image of the instrumental function, or of the motion of cutting. We still do not have an image of the instrumental form, that is, the cutting shape itself of the stone which serves to cut so that the prepared instrument in these conditions is determined solely on the *functional, non-formal* plane. And this is what confirms, as we have seen, the result of excavations in presently known Australanthropi camps, which, although contemporary with *Homo habilis*, correspond in the level of development to the first prehominid phase.

The situation changes from the moment the preparation of the instrument is done at a distance from the object of biological need. In fact, the functional image of which we just spoke, did not find in the simple representation of the biological need object a point of application as firm as the one it already had at its disposal till then in its perception. It therefore had a tendency to fade slowly, so that it had to be made more precise. Thus when, toward the end of the first phase of their development, the Australanthropi, in the absence of the biological need object, indicated to one another various stones which could be used to make a cutting instrument, the simple repeated gesture of the stretched hand was no longer enough to bring them to agreement on a satisfactory choice, since the functional image of the instrument to be prepared, which was still capable of motivating the search for raw material, had meanwhile become too vague to result in something, in other words, too vague for them to recognize among the indicated stones the one best suited for the preparation in question. The simple indicative sign, therefore, had to be completed with a developed indicative sign, probably in the form of a syncretic word, the corresponding gesture being reduced to a simple internal outline. What we obviously are dealing with here is the very sign which was already currently used to designate the instrument in question by the usual movements belonging to it, that is, the movement of its use, indistinguishable from that of its preparation. As the instrument normally appears at rest, the meaning content is constituted here according to formula (2): TFM. On the other hand, there is a question regarding an instrument which does not yet exist, so that the sign can indicate it only as an absent object, according to the representative structure described earlier. We thus have, as a second component of the sign, a developed representative indication whose meaning is 'this here (absent) in the cutting form (C) as it appears in its (presumed) motion', or $\overline{TCM}$. In other words, the stone indicated by the gesture of the outstretched hand is *represented* by the syncretic word as a cutting instrument. The total meaning is thus: 'this here represented as a this here in the cutting form (C) as it appears in its (presumed) motion', or $\overline{T.TCM}$.

Here we meet again the formula of the representation of the absent object (8) with the difference that the simple indicative sign which is its first component now bears on a 'this here present'. On the other hand, in the sign of the developed representative indication, which is its second component, the emphasis in the second place bears on the moment of the form, so that this comes in the second place on the projected image as meaning. And as the 'this here' found in the first place of that meaning is a 'this here absent' and consequently does not contain any singular sensori-motor image; as a result it is the moment of the form, or the cutting shape which in fact appears with the greatest clarity: we thus have here for the first time a representation of the instrumental form. However, as this form is after all only in the second place, and as the sign of this representation continues in close relation with the simple indication of the present material, there necessarily results a certain confusion between the represented image of the instrumental form and the sensori-motor image of the natural stone which the gesture of the outstretched hand indicates as a 'this here'. The whole of that signified thus appears as a syncretic image in which the instrumental form is more or less mixed with the natural contingent traits of the material, and the total sign is defined as the *syncretic sign of representation of the instrumental form*.

Its semantic formula is:

$$\overline{T.TFM}, \qquad (9)$$

'this here represented as a this here in the form (F) as it appears in its (presumed) motion'.

We can also find this sign structure in the child of 16–17 months when he begins to draw.[56] At that age, the child can imitate a line drawn with a pencil on a piece of paper. The line he draws is, of course, not as straight as that drawn by an adult: it is still only a more or less sinuous arc, which approximates the straight line, and is thus clearly distinguished from the scrawl of the chimpanzee, which is a mere sensori-motor exercise. If we analyze the child's gesture, we notice that he begins with a motion of his hand that consists of leaning the pencil on the paper, which is a way of insisting on the movement of the hand stretched toward the object in the indicative sign: '*this here*!' The gesture is followed by a movement which continues to lean on the paper but does so by following a certain form which more or less approximates the straight line. In the projected image, or the sinuous drawing on paper, we can see a certain syncretic confusion between the straight form which the movement of the child's hand tends to impose and contingent characteristics that can accidentally appear on the piece of paper. Thus the whole gesture consists essentially in an association of a simple indication of the 'this here', viz the paper, with a developed indication aiming at a straight line which, in the beginning, does not yet exist. In other words, in this second component of the gesture, we are dealing with a developed representative indication aiming at a 'this here (absent) in the straight line form (S) as it appears in its (presumed) motion', or $\overline{TSM}$. The content signified in its whole is thus: $T.\overline{TSM}$, 'this here represented as a this here in the straight line form (S) as it appears in its (presumed) motion'. We find here again the exact structure defined by formula (9): $T.\overline{TFM}$. And the irregular form of the line drawn by the child is also explained by the syncretic confusion between the sensori-motor image of the present object, namely, the paper indicated as a 'this here' with the represented image of the straight line. As the original instrument, the cutting stone or the stick is characterized in its useful moment by a certain tendency toward the straight line form (or slightly curved), we can consider it as the *instrumental form par excellence*. Consequently, the early drawings of the child of 16–17 months, or about the middle of the prehominid age, appear as a *reactivation of the syncretic sign of representation of the instrumental form*.

The child's sign, however, appears here in a form slightly different from the one described earlier. We have seen that the sign which appeared in phylogenesis was very probably formed by the association of the gesture of the hand stretched toward possible raw material with a syncretic word representing the instrument to be prepared. The gesture of the developed representative indication which gives this word its meaning, reduces itself to a simple internal outline, as is the normal case for an already established word. This signifying form was sufficient for the prehominid group to agree on the choice of the first raw material to be used. Now, when the child begins to draw, the external form of signification is dominated by the gestural aspect. The simple indicative gesture which constitutes its first component, is accomplished in a particularly insistent manner, since the subject pushes on the paper intended as a 'this here!' And this insistence is continued in the second component, that is, the developed representative indication of the straight line, where the gesture of the hand continues to push on the paper, so that the image projected as meaning is actually realized in a drawing more or less approximating the straight line. The total meaning must thus be defined: 'this here represented *with insistence as* a this here in the straight line form (S) as it appears in its (presumed) motion', or by italicizing: $T.\ \overline{TSM}$. By generalization, we obtain the semantic formula of the *insistent* syncretic representation of the instrumental form:

$$T.\ \overline{TFM}, \tag{10}$$

'*this here represented with insistence as a this here in the form (F) as it appears in its (presumed) motion*'.

If we now return once again to phylogenesis, at the time of the transition from the first to the second phase of prehominid development, we may believe that this insistent form of the sign had to appear in the course of the preparation of the instrument, at the time when the situation required a clear representation of the instrumental form. In fact, in the group busy with this operation, an individual can find himself in an perplexing situation, if, for example, while preparing the cutting stone by crushing, he only succeeds in chipping one of the two stones without achieving a useful result. The others then come to his aid. And as they already possess the sign of syncretic representation of the instrumental form, formed at the moment the raw material was chosen, they repeat it, but this time in a more distinct manner, *by emphasizing the gestural aspect*. In other words, one of the companions of the stumped man preparing the tool puts his finger on the chipped but not yet useful stone, and makes a motion which tends to trace the shape of the edge to be obtained at that place on this material. The act of putting the finger on the stone is a way of insisting on the indicative gesture which aims at it as a '*this here*', and the motion which follows marks with insistence the developed indicative gesture which more or less approximately represents the instrument to be prepared in its cutting shape. Here we meet again the very sign of the early drawing by the child in the middle of the prehominid age, and the meaning is defined according to formula (10): 'this here represented with insistence as a this here in the cutting form (C) as it appears in its (presumed) motion', or: $T.\ \overline{TCM}$.

The sign which we have just described in its double form, has appeared only on the objective plane of the *language of real life*, under the pressure of the exigencies of the situation. It constitutes the direct expression of the motion of material activity and of the material relations of the prehominids in the development of their social practice, and such an expression is still part of material behavior itself. However, the very conditions in which the second form of this sign arises entail its cognizance. In fact, as the sign is established in the reciprocal structure, it is sent back from the moment of its reception to the one who emitted it. But as, in fact, the stumped tool-maker. has nothing to say to his companion, since it is precisely up to him to shape the cutting stone that has been indicated to him, the sign which he automatically sends back to the other, *returns upon himself*: it is thus, in fact, for himself that he repeats this representative indication, so that *he addresses it to himself from the other with whom he identifies himself*. In other words, by the insistent gesture of the finger pointing to the chipped stone, the *subject represents to himself* the instrumental form to be shaped from the raw material, which amounts to saying that he *becomes conscious of it*, since at the same time he possesses his own image in the *still present* gesture of the other, identified with himself and thereby appearing in the unity of his own *lived experience*.

We have seen in the first part of our study that in the preparation of the stone instrument by crushing, both of the stones, one held in each hand, functioned together as raw material since the useful effect can be produced from one as well as from the other. When the preparation of the instrument began to be accomplished in the absence of the biological need object, both of these stones take the dominant pole position of the perceptive field, but as they occupy it together, neither one of them can as yet become the principal center of attention. Consequently, their differentiation into instrument of labor and object of labor is not yet possible. The situation changes with the sign just described. When the stumped tool-maker sees the other representing to him, with an insistent gesture of the finger, the shape of the edge to be obtained on the already chipped stone, his attention is necessarily concentrated on it, and he finds himself led to a certain extent to adjust the second stone in terms of the shape that is being represented to him in the first stone. The movement is completed when the stumped tool-maker repeats the sign to himself, which leads him to guide the second stone so as to realize on the first stone this shape which he represents to himself on the first. In other words, cognizance of the instrumental form motivates an activity in which the second stone functions as an *instrument*, the first thereby becoming an *object of labor*.

We are dealing here, of course, with only a sporadic cognizance, which appears on the occasion of individual perplexity. But as any one of them could happen to be more or less stumped, this insistent form of the sign is generalized for the whole group. At a certain time during the preparation of the instrument, the prehominids get into the habit of encouraging one another by each one using his finger to trace the useful shape to be produced on one of his neighbor's two stones. As a result, that motion is integrated into the enduring image of the group which each individual keeps constantly within himself. The sign thus internalized brings the subject *to repeat it to himself without waiting for another to really address it to him*. In other words, from now on, it is *from the beginning* of preparing the instrument, at the moment when he takes the two stones in his hands, that he traces with his finger, for himself, the shape of the edge to be obtained on one of the two stones. And as the hand that makes this sign already holds the second stone, it is obviously the second finger which can most easily straighten out in order to be placed on the first stone. It is thus this finger which will be progressively specialized as the *index finger*. In this individual cognizance, the differentiation of the two stones is achieved, one definitely functioning as object of labor and the other as instrument of labor. Crushing has therefore been transformed into an act of labor: the cutting of the stone by means of striking.

However, since the subject has command of only a *syncretic* representation of the instrumental form, stone cutting, which has just begun, can guarantee only a *summary shaping* or *semi-shaping* of the useful part of the instrument. In fact, as in the representative image of the cutting shape, it remains more or less confused with the contingent aspects of the natural shape of the raw material, the subject being satisfied as soon as he has obtained a more or less useful edge with one or two strokes on one side. And it is the result of this semi-shaping that we can observe on the edge of the Kafuan stone which, cut on one side only, presents an irregular shape where the strictly useful characteristics obtained on the worked surface are mixed helter-sketer with the contingent aspects of the natural side.

F. *Deferred imitation as insistent syncretic sign of representation of the motion of the absent object.* The child's development in the middle of prehominid age is also characterized by the appearance of the deferred imitation phenomenon. Thus in one of Piaget's observations, we see a little girl of 16 months visited by a little boy of 18 months who throws a tantrum. The little girl stands still, watching him in amazement, and the next day it is she who repeats the scene.[57] This child's behavior is obviously related to the scene of the previous day which still remains more or less present by its enduring image, even though it is not localized in the present perceptive field. We thus have first of all, in the outlined form of an internal tension, an indication of the past event as a 'this here absent'. And it is this simple representative indication which the subject specifies by an ensemble of motions which appear as movements of his own body considered as object, and at the same time relating as such to the '*this here absent*'. Thus we have associated with the outlined indicative sign of the 'this here absent', a developed indicative sign defined according to formula (4): MTF, 'the motion of this here in the form (F)'. To which it should be added that this developed indication appears in a particularly insistent form, since in fact in concerns an ensemble of movements which are prolonged for a length of time. This can be emphasized by italicizing the formula: *MTF*. The total meaning can thus be defined: '*this here (absent) representing with insistence in the motion of this here in the form (F)*', or:

$$\bar{\mathrm{T}}.\mathit{MTF}. \qquad (11)$$

Notice that the moment of motion in this formula refers both to the 'this here' present, that is, the subject's own body considered as object, and at the same time to the 'this here absent'. Thus, there necessarily ensues a certain confusion in the semantic content, so that we are in fact dealing with a syncretic representation. Formula (11) thus enables us to define the deferred imitation as the *sign of insistent syncretic representation of the motion of the absent object*.

If we return to phylogenesis, we may believe that this sign had to appear during the transition from the first to the second prehominid stage at the time when preparations were underway for hunting expeditions. As we noted earlier, the subject already had at his disposal the sign of representation of the absent object, that is, gesture of the outstretched hand associated with a syncretic word designating the invisible game in the swamps: $\overline{T}.\overline{TMF}$ (8). In the excitement of the impending hunt, the word became reinforced by externalizing the developed indicative gesture whose internal outline it already implied. As the excitement increased, the gesture was prolonged by mimicking the habitual motion of the game in question. In other words, the developed indicative sign, which constitutes the second component of the total sign, takes on an insistent form and thus appears as a process of imitation. In the image projected as meaning, it is the moment of the motion which now comes in the first place. And as this meaning is effectively realized in the subject's own body, there occurs a syncretic transfer of the indicative meaning which now indicates the motion in question in the subject himself considered as object: this time it is no longer a question of a representative indication but rather a presentative one. This second component of the total sign thus takes on the meaning: 'the motion of this here (the subject's own body) in the form (F), presented in an insistent way', or MTF. And it is this presentation which is now associated with the simple representative indication which aims at the far-off invisible game, so that the total meaning is defined exactly according to formula (11) as the formula of the insistent syncretic representation of the motion of the absent object: $\overline{T}.\mathit{MTF}$.

In short, the Australanthropi hunters, at the time of departure for the hunt, have reinforced their representation of the game by imitating its habitual movements. And it is this insistent image which impelled them to prepare their weapons, for example, clubs, before the expedition itself, which created once more the conditions for the transition from the simple direct manipulation of the raw material to its elaboration by means of a second instrument. It is undoubtedly here that we must search for the origin of *wood* work, since in all likelihood wood originally furnished the most efficient weapons.

The Functional Sentence

We have seen that the new signs which appear in the child in the middle of the prehominid age, at about 16–17 months, allow us to understand better the transition from the first to the second phase of prehominid development in phylogenesis. The second stage of the prehominid age in the child, which goes from 17 to 20 months, must therefore correspond to the second prehominid phase itself, characterized by the development of the elaborated instrument (Kafuan).

In the child, at this stage, we see the first appearance of word associations, often called 'pseudo-sentences' for the reason that, as yet, they do not contain any definite grammatical structure. This denomination, however, does not seem to us to be a very good one, insofar as it suggests that these associations have nothing in common with real sentences except that words are juxtaposed to one another. For we can no longer understand how, under these conditions, the transition from the one to the other could be made. Since in the child's development, these associations occupy an intermediary position between the isolated syncretic words, which are the only ones known between 14 and 17 months, and the first real sentences which begin to be formed at about 21 months, one must indeed believe that they do play a transition role, which assumes that they must already contain some embryonic connections which will be fully formed in the true sentence.

Let us take an example that has been used to show that the subject, at this stage, is only capable of simple word juxtapositions: a baby who called '*little*' all the little pieces of anything, when he wanted a little piece of toast, said, "*baby little*."[58] The relation between the two words is evidently not formulated. Does this mean, however, that there is absolutely nothing in this association which could indicate a syntactical connection?

Since we are dealing here with syncretic words, we must first begin by analyzing their meaning according to the structure of the developed indication. '*Baby*' here designates the baby himself as a 'this here' considered in its motion, according to formula (1): TMF, since the baby is demanding something. Since the child lives entirely dependent on his social environment, we can say that it is the form of demanding which, par excellence, defines the form of his motion, and consequently the moment of the form in the meaning of the syncretic word '*baby*'. In the present case, the baby in question thus indicates himself to himself as the 'this here in a motion in the form of demanding (W)', or: TMW. – 'Little' here, obviously designating the little piece of toast which is at rest, is defined according to formula (2): TFM. The form (F) at stake here, is the form of shortening, and the meaning is: 'this here in the form of shortening (S) as it appears in its (presumed) motion', or: TSM. – As the two developed indicative signs just analyzed mean two different 'this here's' we must distinguish them into T_1 and T_2. The association '*baby little*' can thus be written: $T_1MW - T_2SM$: 'this here$_1$ in the form of demanding – this here$_2$ in the form of shortening as it appears in its (presumed) motion'. We can see immediately that the simple juxtaposition of the two signified images produces between them a relation which expresses the real relation between the baby and the little piece of toast, the first demanding the second. In fact, in the complex image, the motion of the first *this here*, that is the motion of demanding (MW) is applied to the second which thereby appears as its object. This can be made clear by writing:

$$T_1 MW - (T_2 SM),$$

which is stated as follows: 'this here$_1$ in a motion in the form of demanding *concerning* the object – little piece of toast (T_2SM)'.

As the syncretic word gets its entire meaning from the developed indicative gesture which subtends it, whether in an externalized or simply an internal outlined form, it is evident that the semantic connection which emerges here in the association of two words, is explained by the real connection between the two subjacent gestures, which as they rapidly succeed each other, connect themselves thereby in the same dynamic whole. And it is this temporary connection, spontaneously formed on the gestural plane, which will be stabilized later on the verbal plane in the sentences with subject, verb and object: 'baby demands this little piece of toast'.

Thus, if the associations of syncretic words, which appear in the child of 17 to 20 months, do not yet imply the *form* of a sentence, defined by the syntactical connection, they nevertheless perform the fundamental *function* of the sentence, which consists in expressing, by means of sign relations, the relation between things. It would thus be incorrect to consider them as 'pseudo-sentences,' and we propose to call them 'functional sentences'.

A. *The elementary forms of the functional sentence.* If we consider the elementary structure of the functional sentence, as an association of two syncretic words, we immediately notice that it implies three formally possible cases, depending on whether the two words are taken both in their sense of action or in their sense of object, or whether one is taken in its action sense and the other in its object sense. In fact, however, we must exclude the case where they would both be taken in their action sense, for their connection would then have to express a relation between two motions which would be too complex a content for the child at this level. We are thus left with only two cases which may actually occur.

If we begin with the case where the two words are taken, one in its action sense and the other in its object sense, we have four combinations which are formally possible, depending on whether it is a question, on the one hand, of a desired motion or of an observed motion, and on the other hand, of a moving object or of an object at rest:

MFT (desired motion)	$\mathrm{TMF_1}$ (moving object)
MTF (observed motion)	$\mathrm{TF_1M}$ (object at rest)

However, and still for the reason given earlier, we must exclude the case where the word taken in its object sense would indicate a moving object. In other words, we are excluding the following combinations: $\mathrm{MFT - TMF_1}$ and $\mathrm{MTF - TMF_1}$ since they would express, in fact, a relation between two motions, which is practically impossible at the level considered here. We are thus left with only two forms of associations which are actually possible.

Let us first consider the form:

$$\mathrm{MFT - TF_1M}, \tag{12}$$

which is stated: 'the motion in the form (F) concerning this here – this here in the form (F_1) as it appears in its (presumed) motion'.

In Gvozdev's list we find at the age of 20 months "*iço maka*" ("mo milk"). '*Mo*' is a syncretic word which indicates a desired motion in the form of augmentation concerning 'this here'. '*Milk*' indicates the 'this here' in the form of drinking. We thus have as total meaning: 'The motion in the form of augmentation (A) concerning this here – this here in the form of drinking (D) as it appears in its (presumed) motion', or: MAT – TDM, which conforms to formula (12). We immediately see that with the developed indicative gestures which subtend the two words and give them their meaning, a connection occurs which identifies the moment of the 'this here' of the first signified image with that of the second. For the gesture is modeled upon the real situation, and it is, in fact, a question of one and the same 'this here'. We can thus illustrate the connection which is established between the two meanings, by writing:

$$\mathrm{MA\overset{\frown}{T - (T}DM)},$$

which is stated: 'The motion in the form of augmentation concerning this here, namely the object-milk (TDM)'. If we transpose on the verbal plane the connection which has appeared here on the gestural plane, we will obtain the syntactical connection: 'more milk'.

As we remarked at the beginning of this part of our study, the relation of the motion to the object, which we designated by the word 'concerning' is at this level taken in the most general way, since the actual meaning can only be determined depending on the situation. Consequently, in the same form of connection the gesture can project an altogether different semantic synthesis. In '*mo Nini*'[59] for example, the subjacent gesture, after having indicated: 'the motion in the form of augmentation concerning this here' returns to the child who indicates herself: '*Nini*'. The 'this here' which terminates the first meaning is identified with 'Nini', and the relation by which 'the motion in the form of augmentation' concerning the 'this here' is defined is the relation which is projected between the two signified images and which is a relation of attribution. Such a connection, developed on the verbal plane, will give us the syntactical form: 'more for Nini'.

Formula (12) obviously implies its reciprocal:

$$TF_1M - MFT. \qquad (12')$$

Thus in Piaget[60] in the case of a child at the age of 18 months we find: "*ring, ring where it is*?" – "*panene, panana*" (the child says *panana* not only to call her grandfather but also to indicate that she wants something, even when he is not present – her grandfather being the most accommodating instrument of her desires). In Gvozdev we find at 20 months: "*papa, di*" ("*daddy, go*"). In all of these examples, the first word indicates an object at rest: TF_1M, and the second a desired motion concerning it: MFT (the '*where it is*?' evidently functioning as one syncretic word equivalent for 'search'). The gestural connection here implies a ***return of the second indication, at the moment of the 'this here' to the first*** – since the last 'this here' is identified precisely with the indicated object in the first place:

$$(\overset{\downarrow}{T}F_1M) - MF\overset{|}{T},$$

which is stated: 'the object' (TF_1M) in relation with the motion in the form (F) *which concerns it*'. It should be noted that in the first two examples, the 'ring' and '*panene*' are concerned as the object of the motion of searching or asking for. In the third case 'daddy' is concerned as subject of the motion of going expressed by the child's desire. But this difference does not create any problem since at the syncretic level there is a confusion between the subject of the motion and its object.

Let us now go on to the association of a word indicating an observed motion with a word indicating an object at rest:

$$MTF - TF_1M, \qquad (13)$$

'the motion concerning this here in the form (F) – this here in the form

(F_1) as it appears in its (presumed) motion', for example, "*bye-bye daddy*" (Brunet and Lézine), "*bobo patsik*" ("*boo-boo* finger" – Gvozdev). We notice that the first word indicates the motion of the object focussed on (this here) in a certain form: the form of distancing ("*bye-bye*") or the form of painful shudder ("*boo-boo*") or MTF. And the 'this here' is identified with the object at rest indicated by the second word, or TF_1M. The gestural connection can thus be represented by the arrow:

$$MTF - (TF_1M),$$

which is stated: 'the motion concerning this here in the form (F), *that is to say*, concerning the object (TF_1M)'. We thus have the meaning for '*bye-bye daddy*': 'the motion concerning this here in the form of distancing, *that is to say concerning* the object – daddy', and for '*bobo patsik*': 'the motion concerning this here in the form of painful shudder, *that is to say concerning* the object – finger'.

Naturally, when we speak of the object at rest, the notion of 'rest' is used in an altogether relative manner. In '*bye-bye daddy*', for example, we are dealing in fact with daddy who is going, in other words, with an object in motion. But in the analysis effected by the second sign of that functional sentence the object, daddy, is taken not in its motion in the form of distancing, but in another form, from the point of view of being at rest. Since at that age, the child calls every man he meets "daddy" the form in question can be defined by a certain masculine bearing and attire, or form (F_1). From this point of view, 'daddy' is taken in-itself as object at rest: TF_1M. And it is this object which is posited as being concerned by the motion in the form of distancing: "*bye-bye daddy*."

The same external form of verbal association, however, may internally comprise an altogether different gestural connection. A child of 21 months, for example, puts a shell on the table and says "*sitting*." Then she puts it on top of another and says "*sitting on pot*" (Piaget). We can see that we are dealing here with two different 'this here's' so that it is necessary to given them sequence numbers: T_1 and T_2 indicating the two shells. The meaning for 'sitting' is thus: 'the motion concerning this here$_1$ in the form of superposition (S), or MT_1S. 'Pot' designates the 'this here$_2$ in the round form (R) as it appears in its (presumed) motion', or T_2RM. Between these two projected images the following gestural connection is established:

$$\overline{MT_1S} - (T_2RM),$$

which is stated: 'the motion concerning this here_1 in the form of superposition, *is related to* the object – pot (T_2RM)'. Put in syntactical form, this connection will give us: 'sitting on the pot'.

We can see that this form of association, even though externally similar to the preceding one (13), must be defined by a different semantic formula:

$$MT_1F_1 - T_2F_2M. \tag{14}$$

Formulas (13) and (14) have, of course, their reciprocals:

$$TF_1M - MTF \tag{13'}$$

and

$$T_2F_2M - MT_1F_1. \tag{14'}$$

Formula (13′) is applied to functional sentences such as "*daddy gone*," "*boo-boo Nini*" (Brunet and Lézine). The gestural connection is:

$$(TF_1M) - MTF,$$

which is stated: 'the object (TF_1M) in relation to the motion *which concerns it* in the form (F)', namely, the form of distancing for '*gone*', and the form of painful shudder for '*boo-boo*'.

Formula (14′) is applied to functional sentences such as: "*Vonne boom*" (the child hears Yvonne dropping something in the next room – Brunet and Lézine), "*mama soska*" ("*chtchetka*"), broom; the child points to his mother sweeping the floor with a broom – Gvozdev). In each of these two functional sentences we have two different 'this here's': Yvonne and the object she drops, mummy and the broom. As we are dealing here with the reciprocal of formula (14), 'Yvonne' and 'mummy' figure as object (T_2F_2M). '*Boom*' and '*soska*' are taken in their sense of action or MT_1F_1, which stands for '*Boom*': 'the motion of this here_1 in the form of falling', and for '*soska*': 'the motion of this here_2 (the broom) in the form of sweeping'. The gestural connection is:

$$(T_2F_2M) - MT_1F_1$$

or for "Vonne Boom": 'the object – Yvonne in relation with the motion of this here in the form of falling, *which concerns it*'; and for "*mama soska*": 'the object – mummy in relation to the motion of this here (the broom) in the form of sweeping *which concerns it*'. This we can transcribe as follows: "Yvonne, she drops that," and "Mummy, she sweeps with that."

If we now consider the case where the two associated syncretic words are taken both in the sense of object, we will have three formally possible cases, depending on whether the two objects are taken both in motion or at rest, or whether one is taken in motion and the other at rest. But always for the reason given earlier, the first case must be excluded. In other words we are excluding the combination: $T_1MF_1 - T_2MF_2$ since it would still express to a certain extent, a relation between two motions. We are thus left with two forms of association that are actually possible: on the one hand the association between a word indicating an object in its motion, and a word indicating another object at rest: 'this here$_1$ in a motion in the form (F_1) – this here$_2$ in a form (F_2) as it appears in its (presumed) motion', or:

$$T_1MF_1 - T_2F_2M \qquad (15)$$

And on the other hand, the association between two words indicating two objects both at rest: 'this here$_1$ in the form (F_1) as it appears in its (presumed) motion – this here$_2$ in the form (F_2) as it appears in its (presumed) motion', or:

$$T_1F_1M - T_2F_2M. \qquad (16)$$

Formula (15) has already been applied in the functional sentence analyzed at the beginning of this section: "*baby little*." Let us take another example. A child of 17 months who is eating a soft boiled egg holds out a piece of bread saying: "*a pain coco*" (Brunet and Lézine). '*A pain*' is a syncretic word commonly used by children at this age to designate either the object to be eaten or the act of eating it. As the subject here holds out the piece of bread, the word is taken in its sense of object defined according to formula (1): 'this here in a motion in the form of eating (E)', or TME. '*Coco*' can be understood to mean an ovoid or egg-shaped form. As we are dealing here with an object at rest, the meaning is defined according to formula (2): 'this here in the ovoid form (O) as it appears in its (presumed) motion', or TOM. As we are dealing with two different objects, we distinguish them by writing T_1 and T_2, and the total meaning conforms to formula (15): $T_1ME - T_2OM$, 'this here$_1$ in the ovoid form as it appears in its (presumed) motion'. The gestural connection is:

$$T_1\overline{ME} - (T_2OM)$$

'this here$_1$ in a motion in the form of eating *concerning* the object – egg (T_2OM)'. We can see that the motion in the form of eating which is already applied to the 'this here$_1$' can concern 'this here$_2$' only in adjunction to the first. In other words, between the two images projected by the developed indicative gestures which subtend the two syncretic words, a connection in the form of adjunction is constituted, produced by the continuity between the two indications and expressing the real relation between the egg and the piece of bread which the child is eating with it. If we develop this gestural connection into a syntactical form we obtain the sentence: 'this here is eaten with this egg'. We can see that the content of the connection is different from the one analyzed in "*baby little*" even though the form is the same – which is easily understood given the flexibility of the relation that we designate with the word 'concerning'.

Formula (15) evidently implies its reciprocal:

$$T_2F_2M - T_1MF_1 \qquad (15')$$

In Gvozdev's list[61] we find at 21 months: "*dada bada*" (*diadia voda*) = 'uncle' water. The child points to a puddle on the floor made by a family friend who has just shaken the snow off his clothes), "*baba kleça*" (*baba kreslo*) = nanny chair. The child points to his grandmother who is seated on a chair). We see that in these two examples, the first word indicates an object taken in itself as being at rest: T_2F_2M, the form (F_2) being defined according to a certain general bearing of the motion concerning, in the first case, the family friends, and the grandmother in the second case. The second word indicates an object in a motion in the form (F_1): the form of water dripping and running off, and the form of sitting in the chair – or T_1MF_1. Between the two signified images, the following gestural connection is constituted:

$$(T_2F_2M) - T_1MF_1 \quad \text{(arrow from } M \text{ of } T_1MF_1 \text{ to } (T_2F_2M))$$

as for '*dada bada*': "the object – 'uncle' in relation to this here (water) in a motion in the form of dripping, *which concerns it*," and for '*baka kleça*': "the object – nanny in relation to this here (the chair) in a motion in the form of sitting, *which concerns it*." We can approximately transcribe: "'Uncle', that, he made it drip," and "Nanny, that she sat upon."

Finally, if we go to formula (16): $T_1F_1M - T_2F_2M$, we have the connection between two images of objects at rest. For example, in Gvosdev's list we find at 20 months: "*santsik doundou*" (*zaitchik soundouk*, rabbit trunk – The child has just dropped his rabbit behind the trunk). If we designate by

R the syncretic form of the rabbit and by A the syncretic form of the trunk in their (presumed) respective motions,[62] we can establish the meaning of this functional sentence according to formula (16): "this here$_1$ in the form (R) as it appears in its (presumed) motion – this here$_2$ in the form (A) as it appears in its (presumed) motion, or $T_1 RM - T_2 AM$." We see immediately that the motion by which the developed indicative gesture passes from the first object to the second, projects between the two signified images a relation of position which reflects the real spatial relation between the rabbit and the trunk:

$$T_1 RM - (T_2 AM)$$

"the object – rabbit ($T_1 RM$) *in its relation* to the object – trunk ($T_2 AM$)." In syntactical form we will obtain: "the rabbit behind the trunk."

The relation between two objects at rest finds its limiting case in the form of identification. For example, when the child says "*aoua grandfather*" in order to indicate his absence (Pichon) '*aoua*' is defined according to formula (2): 'this here in the form of distancing as it appears in its (presumed) motion'. In other words, he is 'the absent-object', signified in an identity relation with 'the object-grandfather': 'absent-grandfather'. The reciprocal of formula (16) apparently is not distinguished from it externally:

$$T_2 F_2 M - T_1 F_1 M \qquad (16')$$

However, the gestural connection is inverse. For example, in Gvosdev, we find at the age of 21 months: "*niet, maitsik blina*" (no, baby pancake). The child interrupts a question asked by his mother to his father "Do you want a pancake?" The functional sentence '*baby pancake*' here implies a gestural connection parallel to the connection found in the mother's question:

tibié	*priniesti*	*blina*?
(You	want	pancake?)

niet, maitsik	*blina*
(no, baby	pancake)

We can see that because of the situation, the connection here is altogether different from the one found in Pichon's example, "*baby little*." If we designate by B the syncretic form of the baby in its (presumed) motion and by P that of the pancake, we obtain

$$(T_2 BM) - T_1 PM),$$

'the object – baby ($T_2 BM$) in relation to the object – pancake ($T_1 PM$ *which refers back* to him', or 'to baby the pancake'. If we examine the whole of the general structure of the functional sentence as we have just analyzed it, we can make two remarks worthy of interest.

First of all, we can state that for each pair of reciprocal formulas of associated syncretic words, formulas which we have systematically constructed from semantic formulas already acquired from the developed indication, we also have a double figure of gestural connection which is itself reciprocal – the gestural connections being extracted from the analysis of concrete examples. In fact, if we generalize the linking figures which have been presented in each particular case, we obtain the following table:

Direct figures of the gestural connection

1. $MFT - (TF_1 M)$ (12)
   ex: *iço maka* (mo milk) (*mo Nini*)

2. $MTF - (TF_1 M)$ (13)
   ex: *bye-bye daddy bobo patsik* (boo-boo finger)

3. $\overline{MT_1 F_1} - (T_2 F_2 M)$ (14)
   ex: *sitting pot*

4. $T_1 \overline{MF_1} - (T_2 F_2 M)$ (15)
   ex: *baby little* (*à pain coco*)

5. $(T_1 F_1 M) - (T_2 F_2 M)$ (16)
   ex: *santsik doundou* (rabbit trunk)

We can see that in every case the reciprocal figure of the gestural connection is obtained by a simple reversal of the direct figure, corresponding to the reversal of the order of the associated syncretic words – which does indeed verify the *fundamental parallelism of the gesture and of the word.*

Reciprocal figures

1′. $(TF_1M) - MFT$ (12′)

ex: *ring, where it is*
pâtes (pancakes) panana
daddy go

2′. $(TF_1M) - MTF$ (13′)

ex: *daddy gone*
Nini boo-boo

3′. $(T_2F_2M) - \overline{MT_1F_1}$ (14′)

ex: *Vonne boom*
mama soska (mummy broom)

4′. $(T_2F_2M) - T_1\overline{MF_1}$ (15′)

ex: *dada bada* (uncle water)
baba kleça (nanny chair)

5′. $(T_2F_2M) - (T_1F_1M)$ (16′)

ex: *maitsik blina*
(baby pancake)

The second remark that we can make, is that the ten elementary forms of the functional sentence that we have described with the corresponding connecting figures, can be divided into two principal types with one intermediary type. Type I, which includes formulas (12)–(12′) and (13)–(13′), expresses the relation of a motion to an object, which does not essentially go beyond the meaning content of the syncretic word which, used in isolation, already implied such a relation. Consequently, in sentences of this type, we can distinguish a *principal word*, or the word taken in the sense of action, which already expresses the essence of the total meaning, and an *adjunct word*, or the word taken in the sense of object whose role is only to make more precise the moment of the 'this here' in the meaning of the first. In '*mo milk*', for example, the essential is already given in '*mo*': 'the motion in the form of augmentation concerning this here'. And '*milk*' which only makes explicit the 'this here' in question.

On the other hand, type II, which includes formulas (15)–(15′) and (16)–(16′) provides an altogether new semantic content, since it expresses a relation between two objects, either through the mediation of a motion which concerns them both, or directly, both objects being at rest.

Finally, formulas (14)–(14′) belong to an intermediary type, since, on the one hand, they express the relation of a motion to an object, which refers back to type I but, on the other hand, this relation itself already contains to a certain extent a relation between two objects, which refers back to type II.

We can assume that type I, being the most simple, is connected to a more primitive formation: it would only be some sort of excresence of the syncretic word taken in the sense of action, which remains the principal word, the adjunct word making the major word's meaning precise. This does not prevent the two types from appearing almost at the same time, but it is likely that the development, strictly speaking, of type II where the two words are put on the same plane, had to come somewhat later. In little Génia studied by Gvosdev, the two types seem to appear almost concurrently. But obviously the problem can be resolved – only by a large-scale statistical study.

B. *The beginnings of the functional sentence in phylogenesis*. We have seen that the transition from the first to the second phase of prehominid development implied first of all, in the group of Australanthropi hunters, the formation of a vanguard, which profoundly transformed the reciprocal structure of the indicative sign, motivating thereby the transition from the presentative indication to the representative indication. It is probable that the progress brought about in collective work because of the invention of the elaborated instrument led to a new development in the division of tasks, and from this came the reinforcement of the vanguard's role and undoubtedly, on occasion, the more or less temporary appearance of particular teams in the group at work. The framework of communication was therefore considerably broadened and this entailed, in its turn, new progress in the structure of language. It is there, in all likelihood, that we must search for the origin of the functional sentence, whose *reactivation* is seen in the child in the second stage of the prehominid age.

We have seen that the syncretic word used in isolation is comprehensible only as a function of the immediate situation, since it is this which makes possible the recognition of the object, the '*this here*' in question, and the choice between the four available meanings of the developed indication, as they have been presented up to this point in our analysis. This point assumes, of course, that communication takes place between the subjects who work *more or less at the same place*, so that their perceptive fields approximately coincide. Now, such a condition became more and more difficult in the prehominid group as particular teams were formed that could function apart from each other. In other words, the more extensive the field of action of the group, the more complex the coordination of its efforts. By the very play of communication, the ancient forms of language necessarily modeled themselves on the new conditions of collective work, and there resulted the creation of new forms more adapted to the total situation.

Let us suppose that the Australanthropi hunters had to report the flight of the game and thus had to call to one another to pursue it. In the first prehominid phase, when all the members of the group remained more or less together, just one word sufficed, a diffused sound with a meaning comparable to that of "*aoua*" or "*avoua*" of the child understood according to formula (4): 'the motion of this here in the form of distancing (D)', or MTD. In principle, it could just as well mean the distancing of a hunter as well as of the game, but given the situation, doubt was impossible.

But from the beginning of the second prehominid phase, when the use of the elaborated instrument enabled the formation of small distinct teams, the syncretic word used in isolation became equivocal. In fact, the various teams no longer faced the same immediate situation, even when they were close enough to one another to communicate by gesture and voice. Thus in the case before us the diffused word equivalent to the "*aoua*" of the child remained clear for the team that sees the game fleeing, but for a nearby team which is informed only by the sign, it being understood in the representative form $\overline{\text{MTD}}$, it could just as well mean that it was a hunter who was fleeing. A repetition was needed to make things more precise. With the incomprehension of the neighboring team, which does not know how to react, the speaker *repeats* his communication in another form, by adding: "*game*" – it must, of course, be understood that it is once more a diffused sound. For the speaker, *the meaning was the same*. One of the most characteristic motions of the game is fleeing, and in the situation of the speaker it is obviously this motion of flight which is principally aimed at by the developed indicative sign marked with the word '*game*'. It is somewhat the same in the familiar expression: "*minute, papillon*" [just a minute, butterfly], it is mainly the motion of flight which defines the meaning of the word: '*papillon*'. The meaning of the word '*game*' is established here according to formula (4): 'the motion of this here in the form of flight (U)', or MTU, which is a simple redundancy of "*aoua*" as we defined it earlier.

But for the neighboring team, the situation appears altogether different. For it does not see the game in question, and the word which it hears can evoke only the game taken in itself with its full meaning, implying the syncretic form of the ensemble of motions which usually concern it, or the form (G). The meaning is thus defined here in the representative form and according to formula (2): 'this here (absent) in the form (G) as it appears in its (presumed) motion', or $\overline{\text{TGM}}$. Now, the word taken in this sense is normally used to announce the arrival of the game, not its flight. The second team thus looks around expecting to see the animal emerging somewhere. Confronted with such a misunderstanding, the speaker insists: "*aoua game*." Once more he uses not just one but two words redundantly. But this time the second word immediately succeeds the first, and as they have for the receiver two different meanings, he understands them not as a redundancy but as an association: "*aoua game*," or $\overline{\text{MTD}}$ – $\overline{\text{TGM}}$, 'the motion concerning this here (absent) in the form of distancing – this here (absent) in the form (G) as it appears in its (presumed) motion'. We immediately see that the juxtaposition of the two signified images already tends to give rise to a certain identity relation between the two 'this here's'. Because of the reciprocal structure of the linguistic sign in general the receiver *repeats* the communication as he, of course, understood it. The gesture being implied in the word, there is produced a gestural connection which underlines the identity relation that we have just outlined:

$$\overline{\text{MTD}} - (\overline{\text{TGM}}),$$

'the motion concerning this here (absent) in the form of distancing, *that is to say concerning* the object-game ($\overline{\text{TGM}}$)'.

We find here again the gestural connection figure 2 corresponding to formula (13), of the functional sentence such as we have established in the examples: "*bye-bye daddy*," "*bobo patsik*." The addition of the word '*game*' to the word '*aoua*' makes its meaning precise and eliminates the equivocation created by the different situations of the two communicating teams. The new structure is created from the old one, since for the speaker it is merely a redundancy which essentially does not go beyond the already established form of the syncretic words used by themselves. But the receiver understands it as an association of words with different meanings, an association which he repeats with the gestural connection figure outlined in it. In such a *quid pro quo*, in which is reflected the objective contradication between the new relations appearing in collective work, because of the development of the instrumental forces, and the form of language previously acquired, a new linguistic structure is formed for the first time, the *functional sentence* which, even though still composed of syncretic words, nevertheless constitutes a *decisive step* in the progress of knowledge. In fact, the gesture considered by itself is practically limited to the possibilities of the immediate situation: consequently, the gestural synthesis, reduced to its own resources, can at most realize the connection between a simple indication and a developed indication – which we have seen in the preceding section. As soon as it is a matter of connecting two developed indications, the gesture can operate only by being supported by a verbal association, or the functional sentence which marks the beginning of an unlimited development. Afterwards, the verbal synthesis will play an ever increasing role, and it is only by its constant mediation that the subject is able to extend and indefinitely complicate his gestures, in order to *indicate* the external world to *himself* in a more varied and precise projected image.

We have just presented an example of the genesis of a functional sentence of type I. The genesis of type II involves more complex conditions. Let us suppose that a team has just seen the game going around a boulder. In order to communicate that fact to one another and therefore to call one another to pursue the game behind that obstacle, all that was needed for the prehominid hunters, as we have shown in the preceding paragraph, was to stretch the hand toward the rock, in an indicative gesture which aims at the game through that screen, as 'still present' by its enduring image. Another team now appears which has not participated in this last part of the pursuit. The first group of hunters performs the same gesture by adding "*game*" for more precision. For them the word obviously designates the animal in its motion of going around, which it has just done. In fact, one of the characteristic motions of game in general is to disappear behind an obstacle, and in the situation where the first team finds itself, it is that very motion which is particularly aimed at in the developed indicative sign marked by the word. The meaning is thus defined here according to formula (1): 'this here in a motion in the form of going around (A)', or TMA.

But for the second team, things appear differently. For they have not witnessed the preceding scene, and the word '*game*' evokes for them the object-game in its usual syncretic form: 'this here in the usual form of game (G) as it appears in its (presumed) motion', or TGM. For the second team it is a question of a developed indication in the presentative form, since it is accompanied with a gesture of the outstretched hand. But since the hand is stretched toward the rock, the receiver of this communication does not understand at all how the speaker can say "*game*" when pointing to the rock.

Facing this incomprehension, the speaker insists by adding "*rock.*" For him, it is only a matter of a redundancy. For the rock is an obstacle that one can go around. And as the speaker has just seen the game going around that obstacle, it is evidently in that very motion of going around that he indicates the rock. As there is confusion on the syncretic level between the agent of the motion and its passive object, it is clear that for the speaker the word '*rock*' added to the word '*game*' is only repeating exactly the same meaning: 'this here in a motion in the form of going around', TMA.

But once again the receiver hears it differently. For he has not seen this motion, and the word '*rock*' only designates for him the rock itself in its full meaning, that is: 'this here in the usual form of rock (R) as it appears in its (presumed) motion', or TRM. The receiver understands very well now that the hand of the speaker is stretched toward the rock – since it is the rock that is in question here – but he simply does not see what possible interest such a communication can have. Faced with this misunderstanding, the speaker repeats: "*game, rock.*" For him, it is always the same redundancy, but the receiver understands it as an association of words: "*game rock,*" or TGM – TRM, "this here in the usual form of game (G) as it appears in its (presumed) motion – this here in the usual form of rock (R) as it appears in its (presumed) motion'. For the receiver, the gesture of the speaker's outstretched hand obviously always refers to the word '*rock*'. In other words, it is the rock which is actually presented as the 'this here' and consequently, the image signified by the word '*game*' is pushed back outside of the perceptive field, which gives it the meaning of a 'this here absent'. The two 'this heres' are thus clearly distinguished and the total meaning takes the form: $\overline{T_1 GM}$ – $T_2 RM$: 'this here$_1$ (absent) in the usual form of game as it appears in its (presumed) motion – this here$_2$ in the usual form of rock as it appears in its (presumed) motion'.

But virtue of the reciprocal structure of the linguistic sign, the receiver repeats both words as he has just heard them, and he himself stretches his hand in the same direction as that of the speaker. But this gesture now takes on a double meaning: with the word '*game*' it functions as a representative indication since it is a 'this here absent', and with the word '*rock*' it functions as a presentative indication. We immediately see that a gestural connection is being formed which projects a relation of location between the two signified images and represents thereby the real spatial relation between the two objects aimed at: *the game is represented behind the rock*:

$$\overline{(T_1 GM)} - (T_2 RM)$$

We meet here again with gestural connection figure 5 – which we have described in terms of the functional sentence of the child in the example given for formula (16): "*santsik doundou*" (rabbit trunk), except that for the child, the rabbit functions as a 'still present' object, since he has just dropped it behind the trunk, and consequently still keeps its enduring image in that place. The relation between the rabbit and the trunk is thus presented within the perceptive field, and not represented. We have already encountered the same anomaly with the example analyzed earlier, "*aoua game*" which in phylogenesis appears in the representative form whereas the corresponding examples in the child, "*bye-bye daddy*," or "*bobo patsik*" are simply in the presentative form.

Actually the child only operates with a ready-made structure, inherited from the origins of his species, and this is the reason why he immediately applied it within the field of presence of present perception. If we return to phylogenesis, the representative form is altogether essential, for within the perceptive field, the indicative gesture, specified, if need be, by a syncretic word, was sufficient for the workers of one team to call one another for an attack on the present, or 'still present' object. And if another team which finds, or found, itself, far from the field of action of the first, shows its incomprehension, the first team can add a second word, but simply redundantly, since for it the meaning is already perfectly clear. *The functional sentence could thus not originate in the first team, which did not have to go beyond the horizon of the present perception*. It became necessary only for the second team which, not having observed the event with its own eyes, is obliged to 'keep itself informed of the situation', which is possible only by the *explicit representation* of a relation clearly posited between a motion and an object or between two objects. And as such a meaning could not be communicated by the first team, for whom everything is already present, there was a need for the second to construct it itself, and this is what it did on the basis of a misunderstanding. The *quid pro quo* originated in the new objective condition of social labor, from which resulted a new form of language, which gave a richer and more precise image of external reality and which thus answered better the new needs created by the material development of the instrumental forces.

We have just outlined the genesis of the functional sentence as it arose on the objective plane of the *language of real life*, as direct expression of the material activity and material relations of the prehominids at the beginning of the second phase of their development (Kafuan). The process of this genesis, however, already implies a first form of *cognizance*. In fact, the second team which creates the new linguistic structure, has, of course, nothing to communicate to the hunters who have preceded them. It is entirely up to them to become informed of the situation so that they can in fact address the new linguistic sign only to themselves. In other words, the hunters of the second team construct the functional sentence, by indicating to one another and each to himself the objective relation in question, which defines the very structure of cognizance. We are still dealing here, of course, with just a sporadic consciousness. But as the new form of social labor develops, the use of the functional sentence is generalized, so that it ends by being integrated with the enduring image of the group that each keeps within himself. An *individual cognizance* results from this and becomes constantly available to each and every subject.

We have seen that from the beginning the functional sentence implied the representative form: it enabled the teams who find or found themselves distant from the field of present action, to represent to themselves the objective relation which escapes the field of presence of their sense perceptions, and to become thereby 'informed of the situation'. The information obtained in this way, however, still concerns only an invisible relation, in other words an invisible aspect of the present situation. It does not make possible the representation of a totally absent situation.

But once, through the achievement of cognizance, the elementary functional sentence as we have described it, as an association of two syncretic words, has become constantly available to the subject, he can *compose* the representations at his command, in other words, he can associate these elementary groupings of two words so as to construct *developed functional sentences* capable of representing an entire absent situation.

Now, we have seen in the first part of our study that it is precisely the representation of the absent biological need situation which enabled prehominid, toward the end of his development, to employ his leisure moments in the elaboration of the instrument, which led him to pass from *summary* or *confused* shaping to *total* or *distinct* shaping of the useful part of the instrument. And it is this achievement of the work of elaboration which created the premises for the formation of a typical image of the instrumental form, which led to productive labor, with the appearance of *Homo habilis*. We may thus believe that towards the end of the second prehominid phase, the functional sentence was sufficiently developed to evoke the whole of an absent situation, and this is precisely what can be verified in the development of the child where we see evocations of precisely this kind appear toward the end of the prehominid age.

C. *Developed types of the functional sentence*. At about 19 months we can see the child making more than two syncretic word associations which evoke an absent situation. Two principal types are distinguished in them: the enumerative and the correlative.

For the enumerative type we can give Gvosdev's example: at 21 months a child says: "*Lena proua* (to walk), *Tossa proua, kiska* (the cat) *proua*." We can see that we are dealing here with an enumeration by accumulation of associations according to formula (13′), the word taken in the sense of action ("*proua*" = to walk) is constantly repeated. In other words, by composition from an elementary form of the functional sentence a developed form is constituted which to a certain extent evokes the absent situation as a whole; in the example given it is the situation of people walking. "*Léna*," "*Tossa*," "*kiska*" are syncretic words understood in the sense of the object according to formula (2) or: T_1F_1M, T_2F_2M etc. "*Proua*" is defined according to formula (4) as 'the motion of this here in the form of walking (W)', or MTW. We thus have the total meaning:

$$\overline{[T_1F_1M)} - \overline{MT_1W]}, \ \overline{[T_2F_2M)} - \overline{MT_2W]},$$

$$\overline{[T_3F_3M)} - \overline{MT_3W]}, \text{etc.}$$

which is stated 'the object – Léna represented in the motion which concerns it in the form of walking, the object – Tossa represented in the motion which concerns it in the form of walking, etc.'

We have here the developed enumerative functional sentence in its most complete form. In the same enumerative type there are more or less abbreviated forms. For example, one of Piaget's subjects, Jacqueline, at 19 months[63] begins to talk to herself in bed at night in complete darkness: "*Look*: [*tu vois*] *mummy, daddy, grandma, uncle G., etc.*" And she repeats this for a good 10 minutes. Here the word taken in the sense of action is uttered only once (*'Look'* obviously functions as just one syncretic word equivalent to 'there are'). But it is obviously implied before *'daddy'*, *'grandma'*, etc.

The whole has thus, in reality, been constituted by an accumulation of elementary functional sentences according to formula (13). *'Look'*, equivalent to 'there are', can be defined as 'the motion of this here in the form of appearance (A)', or: MTA. *'Mummy'*, *'daddy'*, *'grandma'*, etc., are taken in the sense of the object according to formula (2), or: T_1F_1M, T_2F_2M, T_3F_3M, etc. The total meaning is thus:

$$[\overline{MT_1}A - (\overline{T_1F_1M})], [\overline{MT_2}A - (\overline{T_2F_2M})],$$

$$[\overline{MT_3}A - (\overline{T_3F_3M})], \text{etc.}$$

which is stated: 'the motion concerning this here$_1$ in the form of appearance, that is, concerning the object mummy, the motion concerning this here$_2$ in the form of appearance, that is, concerning the object – daddy, etc.'

However, since the sentence is, in fact, presented in an abbreviated form, the first word being implied from the second association, the actual meaning must be written:

$$[\overline{MT_1}A - (\overline{T_1F_1M})], [\ldots - (\overline{T_2F_2M})],$$

$$[\ldots - (\overline{T_3F_3M})], \text{etc.}$$

which gives us by transfer of all the 'you see' or 'look' (= 'there are') at the beginning of the whole:

$$\overline{MT_{1,2,3}\ldots}A - (\overline{T_1F_1M}), (\overline{T_2F_2M}), (\overline{T_3F_3M}), \ldots$$

The abbreviation can be extended even further. Still at 21 months Génia listens to a story about a dog and to the words: "they gave it to eat"; he concatenates "*maka* (milk), *kissel, kacha, maka, iska* (egg)." We see that the child's utterings actually begin with the last word he heard: "eat . . . ". But since he does not repeat it there results a functional sentence composed solely of a series of words taken in the sense of objects, following formula (2).

Finally, the process is completed when this last form is presented spontaneously. Thus, as mentioned earlier, Piaget's subject getting ready for her nap enumerates the foods just taken "*Ostatine* (*phosphatine, infamil* – infant food formula), *orage* (orange), etc."

We can see that the developed functional sentence of the enumerative type appears in two principal forms. The first consists in a series of elementary associations according to formula (13). The word taken in the sense of action comes in the first place, and afterwards is in general only implicitly repeated so that it is transferred to the beginning of the series: "*Look mummy, daddy, grandma, uncle G.*, etc." The semantic formula is thus:

$$\overline{MT_{1,2,3} \ldots F} - \overline{(T_1 F_1 M)}, \overline{(T_2 F_2 M)}\, \overline{(T_3 F_3 M)}, \ldots \qquad (17)$$

We have a variation of this form, when the word taken in the sense of action is completely implied from the beginning, as in "*Ostatine, orage, etc.*" The formula is then:

$$\ldots - \overline{(T_1 F_1 M)}, \overline{(T_2 F_2 M)}, \overline{(T_3 F_3 M)}, \ldots \qquad (17a)$$

The second form of the enumerative functional sentence consists in a series of elementary associations following formula (13′): '*Lèna walking, Tossa walking, cat walking*'. The semantic formula is thus:

$$[\overline{(T_1 F_1 M)} - \overline{MT_1 F}], [\overline{(T_2 F_2 M)} - \overline{MT_2 F}],$$

$$[\overline{(T_3 F_3 M)} - \overline{MT_3 F}], \text{etc.} \qquad (17')$$

This formula itself contains its abbreviated variation:

$$\overline{(T_1 F_1 M)}\, \overline{(T_2 F_2 M)} \ldots, \overline{MT_{1,2} \ldots D} \qquad (17'a)$$

In Gvozdev, for example we find at 23 months: "*pagui, maki, baï baï* (*sapogui, bachmaki baï baï*: boots, shoes nite-nite).

The enumerative type of the functional sentence does indeed give a certain representation of the absent situation, for example, the surrounding family, the meal, the walk, etc. But we are dealing here with only a vague and incomplete representation, for it only consists in the repeated image of a relation between a certain motion and a series of objects. Actually, the most elementary situation contains at least three well-defined moments: the subject of the situation, its object and the reciprocal dynamic relation which unites them. Its explicit representation thus contains three terms reflecting these three moments, and this is what we find realized in the correlative type.

Piaget's same subject, for example, about ten days after the enumerations mentioned earlier, picks up a blade of grass which she puts in a pail as if it were one of the grasshoppers a little cousin had brought her a few days before. She says: "Totelle [sauterelle, or grasshopper; 'hopper' in what follows], totelle, jump boy [her cousin]."[64] We have here a double symmetrical relation between the middle term and the two extremes. '*Hopper*' and '*boy*' are in effect defined according to formula (2): 'this here$_1$ in the usual syncretic form of the grasshopper (G) as it appears in its (presumed) motion', or T_1GM – and 'this here$_2$ in the usual syncretic form of the boy (B) as it appears in his (presumed) motion', or T_2BM. '*Jump*' is applied partly to the grasshopper's jump, and partly to the sudden motion of the boy's hand which catches it in that jump. Even in adult language the word expresses the two meanings, and it is clear that at the syncretic level both motions are confused in one and the same form of jumping. '*Jump*' is thus defined here, according to formula (4): 'the motion of this here$_1$ and this here$_2$ in the form of jumping (J)', or: $MT_{1,2}J$. The whole is in the representative form since it deals with absent objects, and the meaning of the functional sentence is written as follows:

$$(\overline{T_1GM}) - \overline{MT_{1,2}J} - (\overline{T_2BM}).$$

We immediately see that we have here the abbreviated result of a juxtaposition of two associations defined according to formulas (13′) and (13). In fact, '*hopper jump*' displays exactly the same semantic structure as '*daddy gone*' or '*Nini boo-boo*', and 'jump boy' the same as '*bye-bye daddy*' or '*bobo patsik*'. The gestural connection is thus established according to the corresponding figures 2′ and 2. For '*hopper jump*':

$$(\overline{T_1GM}) - \overline{MT_1J}$$

and for '*jump boy*':

$$\overline{MT_2J} - (\overline{T_2BM})$$

The total figure is thus constituted in a double symmetrical relation:

$$\overline{(T_1 GM)} - \overline{MT_{1,2} J} - \overline{(T_2 BM)}$$

which is stated: 'the object – grasshopper $(\overline{T_1 GM})$ in relation to the motion *which concerns it and which concerns* the object – boy $(T_2 BM)$ in the form of jumping' which we can more or less transcribe as follows: 'the grasshopper it jumps, the boy he catches it'.

The formula of the functional sentence of the correlative type is here written as follows:

$$\overline{(T_1 F_1 M)} - \overline{MT_{1,2} F} - \overline{(T_2 F_2 M)} \qquad (18)$$

We see immediately that this formula is able to express an absent situation since it contains two extremes designating two objects, one of which functions as the subject of the situation and the other as its object, and a middle term indicating the reciprocal dynamic relation between them.

If we return to phylogenesis, we may believe that the developed functional sentence had to appear toward the end of the second prehominid phase, at the time when the instrumental forces reached a level high enough to enable teams of hunters from one group to station themselves at considerable distance from one another for rather long periods of time. If, after some time, they did not encounter each other, then they had to search for each other. And in the tension of that search they evoked one another by accumulating already acquired elementary functional sentences, which amount to a first representation, obviously still incomplete, of the absent situation. For example, each team being usually called by a diffused – syncretic word – either A, B, C, etc. – we would have: "*A gone, B gone, C gone*, etc." Such a structure is reactivated in the sentence of little Génia mentioned earlier: "*Lêna walking, Tossa walking, cat walking,*" etc.

At the same point of the development of the instrumental forces, the role of the vanguard mentioned earlier takes on a new form. We are dealing now with a team of scouts which goes rather far ahead in search of game. They are temporarily separated from the rest of the troop, and we may believe that at the moment of departure there is a recommendation involving the enumeration of the various animals usually hunted by the prehominids, and the listing of this group of animals also represents pretty much the situation of the team of scouts who must look for the animals. The functional sentence is still developed here by an accumulation of already acquired associations, for example: "*Look antelope, look baboon, look giraffe*, etc." Afterward, the first word, taken in the sense of action, is understood to apply to the second association, and we have an enumeration according to formula (17): "*Look antelope, baboon*, etc." Finally, the application of the word taken in the sense of action can be completely implied, particularly if it is a question of little game, for example: "Turtle, lizard, caterpillar, etc." (formula 17a).

As we have already remarked, the enumerative type of developed functional sentence gives only an altogether summary representation of the absent situation. The correlative type, which gives a more explicit representation of it, was probably formed in the development of scenes in which the game was imitated. We have shown in the preceding section that these imitations had to occur while preparing for hunting expeditions, at the time of the transition from the first to the second prehominid phase. We may believe that toward the end of the second phase, because of the progress brought about by the cognizance of the elementary functional sentence, prehominids began to imitate not just the motions of the game but also those of the hunters. There resulted a gestural representation of the characteristic *situation* of the most important moment of the hunt, and this representation was reinforced by the creation of the developed functional sentence of the correlative type.

The functional sentence of Piaget's child, analyzed earlier – "*Hopper, hopper, jump boy*" can be easily transposed to the conditions of phylogenesis. The grasshopper is an edible animal, since it is still being eaten in various parts of the globe, and it is probable that prehominids hunted them. They obviously did not put them in a bucket as Piaget's subject did, but they could have used containers made with big leaves from platan or banana trees. Owing to the cognizance of the elementary functional sentence in the course of the second prehominid phase, the hunters now become aware of the essential relations implied in the situation, since in inner speech they can utter to themselves the elementary functional sentences which represent them: "*grasshopper, jump*" and "*jump hunter*." The imitation scene which precedes the hunting expeditions thus comprises a double motion: the jump of the grasshopper and that of the hunter who catches it. The gesture is reinforced by speech, and there results by juxtaposition and synthesis of the two elementary sentences, the correlative functional sentence: "*grasshopper, jump hunter*" whose reactivation was observed in the child's utterance.

We have just presented the genesis of the two developed types of the functional sentence in the objective dialectic of the language of real life.

However the very conditions of the imitation scenes just spoken of, already imply a first moment of cognizance. In fact, by virtue of the reciprocal structure of the linguistic sign in general, spectators repeat what the actor says. But since they do not have anything to say to him, it is actually *to themselves* that they address that statement, in other words they *take cognizance* of it. On the other hand, we may believe that these scenic representations referred not only to the moment of catching the game, but also to that of the recommendation to the scouts, and as an evocation of the other teams to be sought. Consequently, the cognizance of which we have just spoken also includes the enumerative type of functional sentence.

We are dealing here, of course, with only a *sporadic* consciousness tied to the objective conditions of the preparation for hunting expeditions, in other words to the biological need situation as a whole. But once this new form of scenic representations has been acquired, it could be reproduced at times of *forced leisure*, in bad weather, for example. We are no longer in the biological need situation, but the biological need is always present, which motivates the repetition of these imitation scenes. Since the subjects had nothing else to do, representation was developed. We are no longer dealing now with an actor or with a little team of actors, with the rest of the group as spectators. Time being free, they all participate more or less in the scene so that everybody is at the same time actor and spectator. In other words, everybody tells everybody else and repeats to himself the developed functional sentences which represent the absent situation of the hunt which has not taken place. From this there results a *collective consciousness* which makes available to the group the new linguistic form. In other words, the group ends up by availing itself of it independently of the condition just mentioned, namely the biological need which remains present at times of forced leisure. In other words, the group no longer has any immediate need of that motivation, and henceforth can repeat those scenes even at times of real leisure, when the biological need has already been satisfied.

The representation thus renewed reawakens once again the biological need in the resting hunters and thereby the need to elaborate the instrument. But we are now faced with an altogether original situation. The biological need, when reawakened in these conditions, can appear only in an imaginary form, since the real need is already satisfied. Hence, it can undoubtedly stimulate a work of elaboration, but on the other hand it leaves the subject more or less in a *play* situation.

We know that play has a considerable role in the development of ape behavior. In play new schemes of action emerge which are afterwards applied to the biological need situation. Thus in the famous behavior of fitting two bamboo rods together, experience shows that chimpanzees begin to succeed in that operation while playing, and it is only afterwards that they repeat it in the food-seeking work situation.[65] In fact play makes possible a freedom of movement which arouses all kinds of possible combinations. In the biological need situation, on the contrary, the pressure of the immediate need limits the animal to schemes of action already acquired. In fact as the available nervous energy is strongly polarized on the wanted object, the indirect circuits can open up, in general, only to the extent that they have already been more or less prepared by previous experience, notably the experience of play.

If we return once again to the end of the second phase of prehominid development, we notice that the biological need situation, in the narrow sense, had already been overcome, when the prehominid hunters began to represent hunting scenes in their moments of forced leisure: such a representation impelled the subjects to elaborate the instrument, such elaboration being already outside of the biological need situation, strictly speaking. The biological need, however, is constantly present, and the individual can become quite exasperated in these waiting periods. We may believe that the work of elaboration, performed under these conditions, could not go beyond the usual shape of the elaborated instrument, namely, its more or less confused syncretic shape.

But when, owing to the collective cognizance realized in moments of forced leisure, the prehominid group acquires the use of the developed functional sentence permanently, and thus finds itself capable of repeating the representation of the characteristic situations of the hunt during periods of leisure, it does, indeed, reawaken in itself the biological need, but only in an imaginary form. Now, the imaginary need can not be as engrossing as the real need. Consequently, on the one hand, the subject finds himself impelled to elaborate the instrument which must answer to that need, but, on the other hand, his work is no longer limited by the established schemas.

Once the collective consciousness of the absent biological need situation is acquired, the repetition of imitation scenes ends up by integrating the new linguistic structure in the enduring image of the group. From now on the subject, who has set about the work of elaboration, always keeps within himself the image of those signs which he now repeats in the interiority of his *lived experience*. In other words, the elaboration of the instrument in these leisure moments is accompanied by an *individual cognizance* of the developed functional sentence where the subject continues to represent to himself the absent need situation, so that he constantly entertains within himself the imaginary biological need in question. As a result he finds himself impelled to go beyond the usual, half-fashioned shape of the elaborated instrument. And since his situation contains a certain amount of play, the usual shape no longer constrains him, and he actually goes beyond it by finishing the useful part of the instrument and giving it for the first time a *distinct shape*, produced entirely by labor.

In the first part of our study, we saw that the total shaping of the useful part of the instrument, for example, the cutting of the edge on both sides of the Kafuan stone, was the fundamental condition for the development of the typical shape, which marks the first beginnings of productive labor with the transition from the Kafuan to the Olduvian. We now have to find out by what progress of language and consciousness the prehominid ancestor, starting with the distinct shape he obtained from the work of elaboration, could raise himself to a typical image of the instrumental shape and thus accede to the genus *Homo* as *Homo habilis*.

D. *The disengagement of the form and the birth of the name*. Toward the end of prehominid age we can observe, in the child, developed indicative gestures where, for the first time, insistence on the first place within the formula bears on the moment of the form. Thus in one of Piaget's observations, we see one of his girls, Lucienne, at the age of 16 months,[66] attempting to remove a watch chain from a matchbox which has been opened 3mm. As she cannot succeed in reaching the chain, which has been placed deep in the box, through the small opening, she looks at the slit with great attention; then, several times in succession, she opens and shuts her own mouth, at first slightly, then wider and wider. Finally, she puts her finger in the slit and, instead of trying as before to reach the chain, she pulls so as to enlarge the opening. She succeeds and grasps the chain. The signifying act here contains a double component. First of all, by the direction of her attentive look, the child indicates to herself the slit of the box as the 'this here!' Then by looking at the slit while she opens her mouth, she projects the image of a virtual enlargement of the object. We are dealing here with a developed representative indicative sign, where the insistence in the first place bears on the moment of the form, as the form of enlargement, since the child opens her mouth wider and wider, which brings the form of enlargement to the first level of the signified image. As this form does not yet exist, it is represented as a 'this here absent'. The meaning of this second sign component is thus: 'the form of enlargement (E) of this here (absent) as it appears in its (presumed) motion', or $\overline{\text{ETM}}$. As the child looks at the object with intense attention and repeats her gesture of opening her mouth several times, each time more widely than before, we can consider the whole sign as being executed in the insistent form. The second 'this here' being identified with the first, the total meaning can be written with the gestural connection:

$$\overset{\curvearrowleft}{T.\ \overline{\overline{ETM}}}$$

which is stated: 'this here represented with insistence in the form of *its* enlargement as it appears in its (presumed) motion'. We see that this meaning presents the following structure:

$$\overset{\curvearrowleft}{T.\ \overline{\overline{FTM}}}. \qquad (19)$$

We rediscover here, in the second semantic component, formula (5), FTM, presented earlier, but which only now finds its first practical application.

Formula (19) can be considered as a transformation of formula (10) $T.\ \overline{TFM}$: the moment of the form which came in the second place of the second semantic component, now comes in the first place.

Formula (10) defined the signified content of the child's drawing gesture at 16–17 months, a gesture which we have interpreted as a reactivation of the insistent syncretic representation of the instrumental form, which made possible the elaboration of the instrument in the transition from the first to the second phase of prehominid development in phylogenesis: 'this here represented with insistence as a this here in the form (F) as it appears in its (presumed) motion', $T.\ \overline{TFM}$. We have seen that the position of the form (F) in the second place of the second semantic component ($\overline{\text{TFM}}$) necessarily entailed a certain confusion between the represented image of the instrumental form and the perceptive image of the material indicated as the 'this here!' This confusion determined the syncretic character of the signified whole, witnessed by the syncretic form of the Kafuan instrument. Now, with the transformation accomplished with formula (19) this confusion is averted, since the moment of the form (F) is now distinguished and put in the first place of the second component: $\overline{FTM}$. As a result we have a *distinct* representation of the form, in other words, *for the first time syncretism begins to be transcended*.

This is precisely what we wanted to verify in the child's drawing at about 19 months. As this age the child can already imitate a straight line, without taking the direction into account, and he spontaneously traces lines that are clearly curved to a considerable extent. The progress realized on the sinuous arc which he made at 16–17 months, shows that the gesture has been transformed into a *distinct* representation of the straight or considerably curved form, which implies that the moment of the form has begun to be distinguished by being projected into the first place of the second component of the total meaning. This meaning is thus established according to formula (19): *T. STM*, which is stated: 'this here represented with insistence in the straight or considerably curved form (S) ***which appears as its own*** in its (presumed) motion'.

As we have seen in the preceding section, the straight or considerably curved form can be considered as the instrumental form par excellence. The progress of the child's drawing, from 16 to 19 months, thus appears as a *reactivation* of the development of the sign of representation of the instrumental form which began on the *syncretic* level toward the end of the first prehominid phase, in order to achieve the level of a *distinct* representation toward the end of the second phase. Such a development reflected the material motion of the instrumental forces which resulted, at about the end of the second prehominid phase, in giving the useful part of the instrument a *really distinct* form. In effect, the gesture which draws the shape to be elaborated on the material, for example, the shape of the edge to be cut on the stone, recapitulates the already acquired experience of the shaping of the useful part of the instrument. And owing to the completion of this shaping, the representative gesture itself takes on a completed form, or *T. CTM*, 'this here represented with insistence in the cutting form (C) ***which appears as its own*** in its (presumed) motion'. And this is what we find again in the firmness and clarity of the motion of the child's hand which traces a straight or considerably curved line at about 19 months, in striking contrast with the hesitation displayed at 16–17 months.

The moment of the form here is not yet completely abstracted, since the developed representative indicative sign which projects it to the first place of the formula, is still closely associated with the simple indication of the raw material. The abstraction of the form is completed when the sign in question functions independently, when it is able to enter into new associations. Thus in Piaget's observation about the enumeration of foods, mentioned earlier, we see Jacqueline continuing her discourse with herself, alluding to a newly born cousin. She moved the forefinger of her right hand an inch or so away from her thumb and said, "*Little, little Istine*." It is clear that the gesture emphasizes above all the moment of the form as a form of making something smaller. It is thus this form which is abstracted as the meaning of the projected image which appears in the first position in the formula. The 'this here' is merely outlined and the moment of the motion is simply implied by the fact that the thumb and the index finger are drawn closer together.[67] The meaning of the word *'little'* is thus established according to formula (5): 'The form of making smaller (S) this here (absent) as it appears in its (presumed) motion', or: $\overline{STM}$. *'Istine'* functions as a syncretic word defined according to formula (2): 'this here (absent) in the form of new-born (N) as it appears in its (presumed) motion', or: $\overline{TNM}$. The association of the two words thus constitutes an elementary functional sentence of the new type:

$$\overline{STM} - (\overline{TNM})$$

which is stated: 'the form of making smaller this here (absent) as it appears in its (presumed) motion represented as *belonging* to the object-Istine ($\overline{TNM}$)'.

Since this functional sentence is addressed by the child to herself we can consider its representative form as simply linked to inner speech. If we formulate the structure of the whole, we will thus have to put it in the presentative form:

$$FTM - (TF_1M) \tag{20}$$

which defines the presentation of a *determinate form* (F) as the form of the indicated object (TF_1M).

This formula evidently implies its reciprocal:

$$(TF_1M) - FTM \tag{20'}$$

We can find an example of it in a component of the developed functional sentence in Piaget. Jacqueline at 20 months said: "*mist smoke papa.*"[68] The child was observing the mist on the side of the mountains from her window. "*Smoke papa*" is an allusion to the smoke which covers her father when he smokes his pipe. We can see that the association of three words results from the synthesis of two elementary functional sentences: "*mist smoke*" and "*smoke papa.*" It thus enters in the correlative type of the developed functional sentence already given in the preceding section with the example: "*Hopper jump boy.*" However, here the middle term is not taken in the sense of action, but rather in the sense of the form, namely the wreath-like form characteristic of smoke. The meaning of the word '*smoke*' is thus established here according to formula (5): 'the wreath-like form (V) of this here as it appears in its (presumed) motion', or VTM. '*Mist*' and '*papa*' are taken in the sense of object according to formula (2). Thus if we designate by (B) the syncretic form of the mist and (P) the syncretic form of papa as the child sees them in the usual motion of these objects, we will obtain the following meanings: $T_1 BM$ and $T_2 PM$. We can see that the association '*mist smoke*' is defined according to formula (20′).

$$(T_1 BM) - VT_1 M$$

For '*smoke papa*' the meaning is established according to formula (20), with the representative form for the image of the wreath-form (V) since papa is not smoking at that moment:

$$\overline{VT_2 M} - T_2 PM$$

With the fusion of the two enunciations of the word '*smoke*' in the child's sentence, it is the form of the first (presentative) image which wins out. The total meaning is thus established with the double symmetrical connection:

$$(T_1 BM) - VT_{1,2} M - (T_2 PM)$$

which is stated: 'the object-mist ($T_1 BM$) in the wreath-like form (V) *which belongs to it and belongs to* the object-papa ($T_2 PM$) as it appears in their (presumed) motion'.

Here again we find the reciprocal connection figure presented with formula (18), except that the relation between the two extreme terms is mediated by a form and not by a motion. Thus if we call *dynamic correlation* the double symmetrical relation established by formula (18), we can speak here of a *formal correlation*. We can immediately see that a comparison results from it: the two indicated objects are in fact compared to each other by the mediation of their common wreath-like form. The comparative relation is shown in the relation of T_1 to T_2 in the meaning of the middle term which we symbolize with a colon: $VT_1 :_2 M$. The sentence in question thus takes on the meaning:

$$(T_1 BM) - VT_{1:2}M - (T_2 PM)$$

which is stated: 'the object-mist ($T_1 BM$) presented in the wreath-like form (V) *which belongs to it as it belongs to* the object-papa ($T_2 PM$) as it appears in their (presumed) motion'. In other words: 'this mist is in the wreath-like form *like* papa (when he smokes his pipe)'. – We see that from the abstraction of the image of the form in structures (20′) and (20) their juxtaposition and synthesis constitutes a first comparative structure:

$$(T_1 F_1 M) - FT_{1:2}M - (T_2 F_2 M) \qquad (21)$$

The next day, confronted by the same sight, Jacqueline says: "*mist papa*". This is obviously an abbreviation of the sentence of the previous day: "*mist smoke papa*," the middle term being absorbed by one of the two extremes. Two interpretations are formally possible in this case: "(mist smoke) papa" or "mist (smoke papa)." Probably the first combination is correct, for during the following days she constantly repeats "*cloud papa*" on seeing clouds. In effect, the substitution of '*cloud*' for '*mist*' implies the mediation of a common moment, namely the wreath-like form implied in the meaning of '*smoke*.' In other words, by going from "*mist smoke papa*" to "*mist papa*," the word '*mist*' has absorbed the meaning of the word '*smoke*'. It thus now denotes mist presented in its wreath-like form, and this is what brings it nearer to '*cloud*', whence the phrase '*cloud papa*'.

The correlative abbreviated sentence '*mist papa*' thus has the meaning "*(mist-smoke) papa*" or

$$[(T_1 BM) \cdot VT_{1:2}M] - (T_2 PM)$$

which is stated: 'the object-mist ($T_1 BM$) in *its* wreath-like form as it appears in its (presumed) motion, and *which also belongs to* the object-papa'. In other words: 'this mist in its wreath-like form *is like papa* (when he smokes his pipe).

The word '*mist*' now meaning: 'the mist in its wreath-like form', it follows that its meaning draws nearer to '*cloud*' which enriches its semantic content. '*Cloud*' is first of all an ordinary syncretic word which, taken in the sense of object according to formula (2), signifies: 'this here in the usual syncretic form of cloud (C) as it appears in its (presumed) motion', or TCM. Since the usual syncretic form of the cloud is also in a certain wreath-like form, a confusion with *'mist'* results in cases where that form is also present: there results a semantic transfer in which the word *'cloud'* is enriched with the image that has been added to the syncretic meaning of *'mist'*, or the presentation of the wreath-like form as *determinate form*. This wreath-shape now appears as common to three objects, the mist, the papa and the cloud, and virtually to still others since the transfer can be repeated indefinitely to similar objects. The wreath-like form is thus determined as belonging to an indefinte plurality of *'this here's'* or to a *'this here in general'*, or T_X, which can be identified with various particular 'this here's' without being reduced to any of them. The 'this here in general', of course, can only be represented, since it encompasses a plurality of absent 'this here's'. The word *'cloud'* has thus taken on the meaning: $\dot{T}CM.\overline{VT_XM}$, 'this here in the usual syncretic form of cloud (C) as it appears in its (presumed) motion, and represented in its wreath-like form (V) *which it has in common with other objects*, as it appears in their (presumed) motion.

The association "*cloud-papa*" thus has the meaning:

$$(\overset{\downarrow}{T}_1 CM.\overline{V\overset{|}{T}_X M}) - (\overset{\downarrow}{T}_2 PM)$$

which is stated: 'the object-cloud represented in its wreath-like form *which it has in common with other objects, notably the object-papa*, as it appears in their (presumed) motion'. In other words: 'this cloud which has *the* wreath-like form (in general) is like papa who has it also (when he smokes his pipe)'.

We thus have here a second comparative structure, defined by the following formula:

$$(\overset{\downarrow}{T}_1 F_1 M.\overline{F\overset{|}{T}_X M}) - (\overset{\downarrow}{T}_2 F_2 M) \tag{22}$$

This structure contains two, not three terms, the mediation performed in formula (21) by the middle term which has been internalized in the first, which has thereby been enriched with a new meaning $\overline{FT_XM}$, where the mediating form (F) is in the first place as a form common to various similar objects, or as a form of a *'this here in general'*, T_X. However, the form thus determined as a general form is not yet represented in its abstract generality, but only in its particular realization in the indicated object: T_1. In other words, we do not as yet have a conceptual image of it but simply a *typical* image, since the typical is the general insofar as it is realized in the particular.

On the other hand this typical image itself does not yet appear distinctly since in the whole, signified by the word: $(\bar{T}_1 F_1 M . \overline{FT_X M})$, the general representation $\overline{FT_X M}$ is closely associated with the syncretic image $T_1 F_1 M$, so that there necessarily results a certain confusion between the general form $(\overline{FT_X})$ and the syncretic form (F_1). For *'cloud'*, for example, $(T_1 CM . \overline{VT_X M})$, the determinate wreath-like form as general form $(\overline{VT_X})$ is more or less confused with the syncretic form of clouds (C) as it appears to the child in the habitual motion of that object. Thus we are still dealing here with a *confused typical image*. In other words, syncretism has not yet been transcended. And we can see in the very formula of the word $(\bar{T}_1 F_1 M . \overline{FT_X M})$, that the typical form, or the general form (FT_X) as realized in the particular object T_1, will be clearly abstracted only when the syncretic form (F_1) becomes recessive so that the form (F) will appear *practically* as the only one to be considered. This is precisely what happens in the case of the dialectic of the *name*.

At the end of the prehominid age, we observe in the child of about 20 months a characteristic behavior which brings about the transition to the following stage: the child in her monologues seeks to name things by asking "*what is that*? [*qu'est-ce que c'est, ça*?]" We are dealing, of course, with one and the same syncretic word "*what's that*? [*kèçèça*?]" whose meaning is defined by the questioning gesture projected on the already acquired semantic structure. The child points the finger to the object in question and turns his eyes toward the observer with a look which tends to move alternatively from the observer to the indicated object, and expresses in this way his expectation of that object's name. At this level, the child already possesses the comparative structure. And as the expected name must bring the maximum possible information at his level, we may believe that it must present the semantic structure defined in the first member of formula (22) since it is the richest obtained so far for just one word: $(\bar{T}_1 F_1 M . \overline{FT_X M})$. In this structure, in which expectancy defines the child's question, the questioning cannot, of course, bear on the first component $T_1 F_1 M$ since this deals only with the meaning of a developed indication for which an ordinary syncretic word would suffice. From the very beginning of the stage under investigation, the child himself invents words of this kind: he does not have to ask for them. The questioning which appears at about 20 months can thus only aim at the second component of the meaning in question: $\overline{FT_X M}$. As a matter of fact, we are dealing here with the representation of a determinate form as common to an indefinite number of possible objects. Such a representation implies a stability resulting from previous comparisons and contrasts with the variability of the syncretic image $T_1 F_1 M$ where the moment of the form (F_1) can be modified more or less according to the needs and configuration of the present situation. Hence the expected *name* here must have a fixity which the syncretic word did not have since the child invented and changed it according to the circumstances. Now, the fixity of the name can be obtained only by a social agreement since it corresponds to the experience of a multiplicity of objects. This total experience transcends the immediate possibilities of the development of the present situation and this is the reason why the child asks for the names of things.

In short, what is questioned in the '*what's that*?' of the child which the expected name must determine, at least virtually if not really, is the form (F) of T_X, which, through the mediation of T_X, must be affirmed of the object T_1, which is already indicated according to the ordinary syncretic structure T_1F_1M. The meaning of '*what's that*?' can thus be written as follows:

$$\overset{\downarrow}{T_1}\,F_1M.\,\overline{F?T_XM} \qquad (23)$$

The question mark placed after the F gets its meaning from the questioning gesture mentioned earlier. The motion of the finger pointed toward the object indicates it as a certain 'this here', T_1, which the child can define to himself by a syncretic form, or T_1F_1M. The look which turns toward the observer and tends to go to and fro between him and the indicated object, awaits the determination of form (F), in order to realize the general representation whose formal structure the child already possesses: $\overline{F?T_XM}$. The whole of formula (23) can thus be stated: 'this object indicated in the syncretic form (F_1) is in what general form?'

The name given to the questioning child, is, of course, understood by him within the framework of the question itself, in other words, it takes the semantic structure in abeyance: ($\overset{\downarrow}{T_1}F_1M.\overline{FT_XM}$). As shown earlier, such a meaning is defined as a *confused typical image*, since the general form ($\overline{FT_X}$), represented as realized in the particular object T_1, is still more or less confused with the syncretic form (F_1). When, however, by virtue of the reciprocal structure of the linguistic sign in general, the child repeats the name by outlining the gestural ensemble which it implies and which gives it its meaning, he has a tendency to insist on the second component, $\overline{FT_XM}$, since it is this representation which answers to the question "*what's that*?" Consequently, in the first component, T_1F_1M, the syncretic form (F_1) tends to disappear. In the give and take of questions and answers, the semantic structure of the name is transformed in such a manner that its syncretic content F_1M gradually becomes recessive, while its determinate content $\overline{FT_XM}$ becomes progressively dominant. The transformation is complete when the subject asks himself the question in order to answer it himself.

Thus in one of Piaget's observations, we see Jacqueline at 21 months all alone saying to herself: "*What's that Jacqueline? what's that? ... There* (knocking down a block). *What's falling? A block.*" It is obvious that what interests the child here, from beginning to end, is not the particular block, of which she already has a more or less unstable syncretic image obtained by a developed indicative sign which varies according to the situation, but the block insofar as it realizes the form of the manipulable thing represented in a stable way as common to all blocks in general. She asks herself the question and answers it herself precisely in order to fix within herself that recently acquired meaning, or as we say, in order to 'get it into her head'. In the semantic content of the name, the stable representation of the form determined as such and such a form in general thus definitively becomes dominant, while the syncretic form directly projected by the developed indicative gesture becomes recessive. We can formulate this by writing F_1M in small letters:

$$(T_1 f_1 m.\,\overline{FT_XM}) \qquad (24)$$

We have here the structure of the typical name, as the indication of the object in its distinct *typical form*, as exemplary *type*. The typical is not yet the general as such, but the general in the particular. Its determination obviously already constitutes a first form of abstraction, since it implies the recession of the syncretic content f_1m closely linked to the contingent traits of the object and of the situation, but it is not yet the abstraction of the concept. In fact, this syncretic image, though recessive, still remains, and we see in formula (24) that the general representation $\overline{FT_XM}$ is not yet abstracted from it, so that the general form FT_X does not yet appear in itself as an abstract character defining a class of objects, but only insofar as it is realized concretely in the indicated object. A certain confusion is thus still present between the particular and the universal, a confusion which in the child is prolonged to a rather advanced age. Thus in one of Piaget's observations, Jacqueline, when two and a half years old, used the term '*the slug*' for the slugs she saw every morning on the same road. On seeing one she cried, "*There it is*," and when she saw another ten yards further on she said "*There's the slug again.*" The observer tries in vain to make her distinguish it from the preceding one. As Piaget remarks, for the child slugs are always "*the slug*" reappearing under various forms. It is "a kind of typical individual reproduced in several copies."[69]

At the stage at which we have now arrived, the indication of the object as exemplary type, by its *typical denomination*, marks a decisive step in the progress of knowledge. *For the first time, syncretism is overcome, undoubtedly partially, but actually*: the general form is defined even though only in its particular realization. And we have seen that, in phylogenesis, it is precisely the constitution of this typical image which enabled the prehominid ancestor toward the end of his development to take his first steps toward the labor of production, which raised him to the level of humanity in the form of *Homo habilis*.

We have shown that the first abstraction of the image of the form, obtained by structure (19), $\text{T}.\ \overline{\text{FTM}}$, had its origin in the real abstraction of the instrumental form, when toward the end of the second prehominid phase, the completion of the shaping of the raw material gave the useful part of the instrument an actually distinct configuration. As a consequence, during the course of labor in a later period, the drawing gesture, which projects this form on the material to be worked, assumes a clarity and firmness which we again encounter in the straight or largely curved line traced by the child towards the end of the prehominid age. In other words, in the structure of the representative indication implied in that gesture, $\overline{\text{FTM}}$, the insistence in the first place of the formula now bears on the moment of the form (F). Now, once acquired, this new signifying figure no longer needs to be repeated on the material itself, but may be repeated at a certain distance from it. In other words, the sender of the sign no longer has to come to the side of the clumsy worker in order to put his finger on the stone and draw on it the form to be shaped. By remaining in his place, he can draw the line in question in the air and point to the other's stone in order to tell him what to do. Thus we have now two successive and clearly distinct signs: first, a representative indicative gesture which draws the cutting shape of the instrument to be elaborated in the air, used with the corresponding diffuse-syncretic word which we can transcribe by '*cutting*', or $\overline{\text{CTM}}$; secondly, a developed indicative sign externalized by another diffuse-syncretic word transcribed by '*stone*', taken in the sense of an object: 'this here in the syncretic form of stone (S) as it appears in its (presumed) motion', or TSM. We now have an elementary functional sentence, '*stone cutting*', which means:

$$\overline{\text{CTM}} - (\text{TSM})$$

'the cutting form (C) of this here (absent) as it appears in its (presumed) motion, ***represented with insistence on*** the object-stone (TSM)' – or: 'In the cutting form, this stone!'

Here we meet once more with structure (20) analyzed earlier in the case of Piaget's child who said: "*Little, little Istine*" while also executing in the air the determinate sign of representation of the form of making smaller by drawing the thumb and index finger a few centimeters closer together. The sole difference is that originally it was a question of a call intending to assure the usual quality of collective labor, which implies an imperative insistence, whereas the child is only trying to master a structure inherited from the ancestral past, so that his sentence is uttered in the indicative mood while keeping something of the primitive insistence by repeating the word '*little*'.

Afterwards, the sign of the determinate representation of the instrumental form could have intervened in the choice of the raw material. Prehominid workers indicate to one another various stones more or less suitable for shaping by saying "*cutting stone*." We have the inverse sentence of the preceding one, in a less insistent tone, the representative gesture of the cutting shape being simply outlined:

$$(\text{TSM}) - \text{C}\overline{\text{TM}}$$

Again we meet with structure (20′) with the second term in the representative form, or approximately homologous to '*mist smoke*' which constituted the first component of the comparative functional sentence of Piaget's child: "*mist smoke papa*."

At the moment when they set to work, the Australanthropi, as they evolved from the end of the Kafuan period, could encourage one another by repeating the same sentence "*cutting stone*" in an imperative tone which demands the cutting shape. The meaning is then written:

$$(\text{TSM}) - \text{C}\overline{\text{TM}}$$

which is stated: 'the object-stone (TSM) represented with insistence in the cutting form (C) *which must belong to it*, as it appears in its (presumed) motion'. In other words: "This stone, in the cutting form!"

In the Australanthropi, the mature males, who join experience with strength, undoubtedly had much prestige and authority, as we may still observe in the anthropoids. Thus it was after them that the young patterned themselves when the prehominid group set itself to the elaboration of instruments. These elders were probably called in a particular way: a diffuse-syncretic word, which we can transcribe with the word that the child uses to call adults of a masculine bearing: 'daddy'. In the shaping of stones, the young workers who were stumped could thus indicate a model to one another in the instrument elaborated by an experienced hunter by saying: "*cutting daddy*." The word '*cutting*' being first taken in its sense of object: 'this here in the cutting form (C) as it appears in its (presumed) motion', or TCM. But with the acquisition of the sign of determinate representation of the instrumental form, there was no longer any need to indicate the cutting-model itself. A drawing gesture in the air, which we have just described, with the word '*cutting*' followed by '*daddy*' was sufficient for the purpose. '*Cutting*' is now taken in the sense of 'the cutting form (C) of this here as it appears in its (presumed) motion', or CTM. The meaning remains presentative since the cutting-model of 'daddy' is always there. '*Daddy*' keeps its sense of object: 'this here in the usual syncretic form of the daddy (D) as it appears in its (presumed) motion', or TDM. '*Cutting daddy*' is thus now a functional sentence of structure (20):

$$CTM - (TDM)$$

which is stated: 'the cutting form of this here as it appears in its (presumed) motion, presented as *belonging to* daddy (when he shapes his stone)' – just as in Piaget's child, the expression "*smoke papa*" had appeared with the meaning: the wreath-like form presented as *belonging to* daddy (when he smokes his pipe). '*Cutting daddy*' can thus be more or less translated as: 'the cutting form of the stone of daddy'. In passing, we should note that the confusion produced here between 'daddy' and 'the stone of daddy' is altogether normal at the syncretic level. Thus in Gvozdev's list we see little Génia using the same word '*tota*' ('*tiotka*' – aunt) to indicate either his aunt, or her basket.

The functional sentence just acquired might have been associated later on with the one the hunters were already accustomed to utter when preparing for the task of shaping the stone: '*stone cutting*'. In fact, if, in the course of shaping a stone, a young Australanthropus found himself stumped, the others could encourage him by repeating, "*stone cutting*," and by adding for precision, "*cutting daddy*."

The total meaning is written by quantifying the 'this here' as T_1 and T_2 in order to distinguish the stone of the young worker from 'daddy':

$$[(T_1\,SM) - \mathit{CT_1 M}] - [\overline{CT_2}M - (T_2\,DM)],$$

in other words: "This stone, in the cutting form! The cutting form of daddy's stone."

Afterwards, these two associated sentences may have been repeated at the end of the shaping task in order to indicate the satisfaction of the young workers about the result of their labor. The first sentence then becomes simply presentative of the shaped stone in the cutting form, and the total meaning is:

$$[(T_1\,SM) - CT_1 M] - [CT_2 M - (T_2\,DM)],$$

which can be translated as: 'This stone is in the cutting form, the cutting form of daddy's stone'.

The two enunciations of the word '*cutting*', having now applied exactly the same semantic content to both stones, fuse together and the two sentences are reduced to one: '*stone cutting daddy*', which is the exact homologue of the comparative sentence of Piaget's child: '*mist smoke papa*'. The meaning is thus established in the same manner according to formula (21):

$$(T_1\,SM) - CT_1 :_2 M - (T_2\,PM)$$

which is stated: 'the object-stone (T_1 SM) presented in the cutting form (C) *which belongs to it as it belongs to* the object-papa (T_1 PM) as they appear in their (presumed) motion'. Or: 'This stone is as cutting as that of papa'.

Once this comparative structure is acquired, it can be abbreviated by leaving out the middle term, since in the comparison the emphasis is on the two extremes. The subject of the comparison or the stone of the young hunter absorbs the image of the cutting form, since, in fact, it has just taken on precisely that form, and it is this stone in its cutting form which is being compared to the elder's model. We thus have the abbreviated sentence '*stone papa*', in the sense of '(*stone-cutting*) *papa*', which is homologous to '*mist papa*', of Piaget's child. The meaning is therefore analogous:

$$[(T_1\,SM).\,CT_1 :_2 M] - (T_2\,PM),$$

In other words: 'this stone in its cutting shape is like that of daddy'.

This comparison is repeated by the young Australanthropi as they look one by one at the different stones they have just shaped. And as they award themselves this satisfaction at the end of every work session, the image of the cutting shape is progressively referred to an indefinite plurality of examples. This image thus tends to be distinguished from the contingent traits of the various particular situations and thereby becomes more and more representative of the cutting shape of cut stones in general. Because it is repeated the comparison in question tends to assume structure (22) defined earlier following the child's phrase: "*cloud papa*" or:

$$(T_1\,SM.CT_XM) - (T_2PM)$$

We know that at the syncretic level the same object can easily change names. In one of Konnikova's observations, for example, a boy of 17 months calls his wooden horse "*no*" then "*liaka*." When '*liaka*' was thus substituted for '*no*', it almost immediately took the place of '*o-o-o*' which until then was applied to the automobile, and of '*din-din*' which was applied to the streetcar.[70] We can see that the name change corresponds to the appearance of a new form of the subjacent gesture of the developed indication, which now also encompasses the form of the car and that of the streetcar as the child sees these objects in their usual movement.

We may thus believe that in the case of young Australanthropi at about the end of the Kafuan, the new meaning, $(T_1\,SM.CT_XM)$, which tended to appear in the repeated motion of comparisons ended up by becoming fixed in a new diffuse-syncretic word. We may transcribe this word by '*chopper*', the chopper being defined here as the shaped stone represented in its cutting form *which it is has in common with other similar ones*. Thus we have now the new comparative sentence, '*chopper daddy*' which is homologous to the child's sentence, '*cloud daddy*'. The meaning, already established earlier according to formula (22) can be transcribed: 'This chopper which has the cutting form *like the others*, is like that of daddy', or: 'This chopper which has *the* cutting shape (in general) is like that of daddy'.

We have here a new structure of the word, which still remains syncretic, but which already shows signs of differentiation. In fact, if we examine the meaning of 'chopper' as it has just been constituted $(T_1\,SM.\mathit{CT_XM})$, we see immediately that the presence of the general representation CT_XM tends to prevent the syncretic meaning $T_1\,SM$ from changing into another variation of the developed indication. For the cutting shape (C) now appears as a general form realized in the particular object T_1, in other words as a more or less typical form, which in its particular configuration refers implicitly to an indefinite plurality of possible objects. Consequently, the whole image of the shaped stone, signified by the word, '*chopper*', no longer slides, so to speak, into a sense of action or object in motion, unless for some reason or other, the syncretic component $T_1 F_1 M$ takes over, in which case the word '*chopper*' tends to become an ordinary syncretic word again like '*stone*' as we have defined it, which would enable the moment of motion to come in first or second place in the formula. In other words the word '*chopper*' tends to become specialized in the sense of an object at rest, which means that it *begins to be differentiated as a substantive*.

The new structure, which appeared in the development of the elaborative work on the instrument, is extended to various objects appearing throughout the whole period of prehominid labor. New words are thus constituted which indicate these objects to a certain extent in their more or less typical form. This progress can be found again in the sudden vocabulary development occurring in the child at about 19–20 months. The new words tend to escape the polysemy of ordinary syncretic words, since, on the one hand, they are ordinarily taken in the sense of object at rest, and, on the other hand, the form (F) presented in its more or less typical image no longer has the instability of the moment of the form signified by the ordinary syncretic word. We have seen, in fact, that the form in which the syncretic word indicates the object with its movement is variable according to the situation. Thus '*daddy*' may designate men lighting their pipes, those who hold out their arms to children, those who appear at the window, etc. (Piaget). '*Papo*' may refer to any form of motion or objects concerning walking (Bourjade). '*Tota*' sometimes designates the aunt, other times her basket (Gvozdev). With the new structure as it results from the comparisons of objects through the mediation of a common form, this form takes on a more or less typical figure of a form which relates the word to the thing independently of the immediate situation. We can thus consider this word as a *name*, since it designates the thing *in itself*. However, we are still dealing with nothing but a *functional name*. As a matter of fact, we have already remarked that in its semantic formula $(\overline{T_1 F_1 M. FT}_X M)$, the typical form (F) still remains in a certain state of confusion with the syncretic form (F_1). Actually, then, we have not yet gone beyond the syncretic level, and with every indicated object the typical image must be reconstituted in a more or less confused manner. In short, the functional name still indicates the object in its *confused typical form* only. The presence of this new content, however, is enough to give rise to a problem which was previously non-existent.

Let us suppose that the Australanthropi hunters of the end of the Kafuan perceive an object at a distance which resembles a tree as much as a baboon. Before, in such a situation they could use a developed indicative sign, which, because of the instability of the signified syncretic image, intended the object indifferently either in the form of the usual movement of the baboon, or in that of the tree: everything depended on the mood of the subject at that moment, since he did not have any higher criterion at his disposal. Thus in one of Piaget's observations, Jacqueline at 18 months said "*frog*" when looking at a mark on the wall: influenced by another form she could just as well have indicated that mark by another word. But now that the new structure of the functional name is established, the Australanthropi had at their disposal determinate names for the baboon and the tree, since they are objects encountered in daily work. And since they imply a more or less typical image of the form of baboon (B) and that of the tree (A) the subject may no longer use them indifferently in the equivocal situation in which he finds himself. It is true that it is still a question of merely confused typical images. But given the seriousness of the biological situation, they tend to exclude one another since they call for different behaviors. And as the subject cannot choose between them, he is forced to utter both names one after the other. We are dealing here, of course, with two diffuse sounds which we transcribe as '*baboon*' with the meaning:

$$(\overline{T_1 F_1 M}.\, BT_X M)$$

and '*tree*' (*arbre*):

$$(T_1 F_1 M.\, A\overline{T_X M})$$

The first semantic component is written in the same manner in both meanings, for the individual differences which may be introduced in the syncretic image of form (F_1) are without importance, since the syncretic forms easily pass into one another. But the typical represented forms, even confusedly, require a choice: A or B, '*tree*' or '*baboon*'. And as the hunters are still unable to choose, they look at each other, then once again turn their eyes toward the object while repeating both names in question. In this alternating movement of the look accompanied by the repetition of the names, a new meaning is constituted. In effect, the gesture and the word tend to be in unison, or the succession of the two names tends to be in rhythm with the movement of the eyes. The subject thus concludes by saying, for example: "*tree*" while looking at the object, and "*baboon*" while looking at his companions. From this results a parallel oscillation between the two images projected as meaning:

$$(T_1 F_1 M . \overline{A T_X} M) \rightleftarrows (T_1 F_1 M . \overline{B T_X} M).$$

We can see immediately that in these two alternating images the identical moments combine, so that the oscillation continues only between the more or less typical forms of the tree and of the baboon, A and B:

$$T_1 F_1 M . \overline{(A \rightleftarrows B)\, T_X} M.$$

This oscillation, which is projected between the signified forms A and B, defines the meaning of the signifying motion as *questioning gesture*. Afterwards, the structure which has just appeared may be extended to other relatively typical forms, of A (tree), B (baboon), G (giraffe), etc. The alternating motion of the signifying then tends to be projected in a circular figure:

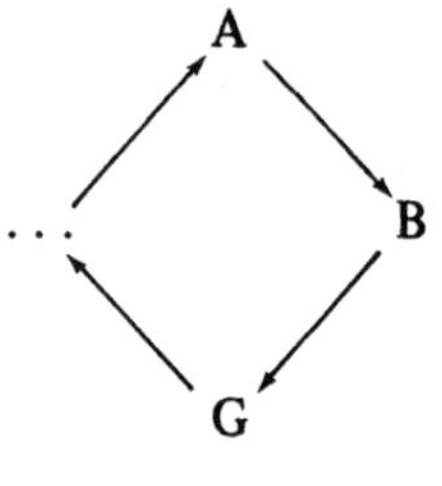

Since it is this very same figure which is intuitively symbolized by the question mark (?), we can write the total meaning:

$$T_1 F_1 M\, \overline{(A?B?G? \ldots ?)\, T_X} M.$$

Finally, in situations which are completely equivocal, the questioning gesture tends to make the rounds of all of the more or less typical forms which the subject has at his disposal in his representation and which we designate as a whole by 'F'. The interrogative meaning is then written:

$$T_1 F_1 M . \overline{F?} T_X M$$

Here again we meet with formula (23) with which we had defined the '*what's that*?' which appears in children's speech at the end of prehominid age, during the transition to the following stage. We may believe that in phylogenesis the constitution of the interrogative gesture at the end of the Kafuan was also accomplished by the formation of a new word, a diffuse sound which we also transcribe by '*what's that*?' and which brings about the transition between the functional name and the typical name, as the first differentiated word.

As a matter of fact, once constituted, the '*what's that*?' could occur at the end of the instrumont elaboration sessions, where we saw the young Australanthropi expressing their satisfaction by comparing their shaped stones with those of the elders and saying: "*chopper daddy*." This scene could change, for once the functional name '*chopper*' is acquired, the reference to '*daddy's*' model is virtually transcended. During and after the instrumental shaping session, the young workers could, if necessary, simply refer to the relatively typical image of the cutting shape evoked by that name. And this is precisely what happened when they had to cut their stones in the absence of their elders. Actually, they had no other resource in these circumstances than to evoke the model of the cutting edge by saying: "*chopper*." And once the instrument was shaped, they repeated the same name to confirm their success. However, as in the total image evoked by that name, which guided the elaborative work ($\overline{T_1 SM.CT}_X M$), there remains a certain confusion between the typical image of the cutting form (C) and the syncretic image of the form of the stone (S), since '*chopper*' is still only a functional name; as a result, the represented model with which the young workers had to be content here is not as clear as the one they had in their ordinary work sessions where they were guided by the same image which was also concretely related to the choppers of their elders. Thus cases may occur where shaping does not quite succeed, so that it becomes difficult to recognize in the completed work the more or less typical form that was supposed to be realized. The subject thus cannot feel the satisfaction described earlier, in other words he cannot *name* the pseudo-instrument just produced. And as he is perplexed, his neighbor intervenes with the word for equivocal situations of this kind that is already in use by the Australanthropi of the end of the Kafuan: "*What's that*?" The clumsy worker can only answer by repeating to himself sheepishly, "*what's that*?". He thus begins his work over again and finally obtains a suitable cutting instrument. But still affected by his previous failure he repeats once again the same question addressed shyly to the same neighbor: "*what's that*?" And the other, recognizing the instrument, shakes his head: "*Chopper*!" The subject repeats this name with satisfaction, and in the joy of success, he repeats for himself both the question and the answer: "*what's that*? . . . *Chopper*!"

We have here a first *cognizance of the instrument* in its instrumental shape determined as *typical form*, in an idealization which presents it in a distinct manner for the first time. In fact, as the subject carries on an inner monologue resulting from the enduring image of the neighbor who originally asked the question and answered it, he at the same time has his own image in this image of the other with which he identifies in his own *lived experience*. Now, this very process of cognizance clearly distinguishes the typical image implied in the meaning of the name. In fact, since the question that the subject repeats to himself refers precisely to the cutting form (C) determined as general form, it is the representation of that form $\overline{CT_X M}$ that he emphasizes in the gestural ensemble subjacent to the name by which he answers himself, so that the syncretic form of the stone (S) passes to the recessive state in the formula. The meaning of '*chopper*' thus now assumes structure (24):

$$(T_1\,sm.\,\overline{CT_X M})$$

The syncretic content of the image of the stone (sm) can now be considered as more or less negligible, and it is clearly the cutting shape determined as general form realized in this particular stone which the subject indicates to himself *distinctly in its typical form*. We should note that this distinct typical form does not yet really exist in that shaped stone, since in the work of shaping, the worker commands the functional name '*chopper*' for only a confused typical image. The cutting edge thus obtained on the real chopper could appear only as a relatively regular form. But by virtue of the cognizance just described, this still slightly irregular chopper is *ideally* indicated in its distinct typical form, or in the image of a completely regular cutter. In fact, in the lived motion of the dialogue with himself, because of the intentionality of the '*what's that*?' described earlier, the subject *neglects* the syncretic content ($f_1 m$) which becomes recessive, so that he neglects the actual irregularities of the real cutter reflected in this syncretic content in order to point out the instrument to himself distinctly in its *ideally typical* cutting shape.

The cognizance just described is still *sporadic* since it is linked to the particular situation of a young clumsy worker. But as it may happen that everyone at some time or other may fail in his work, the newly arisen structure is generalized. And at a certain point the hunters, after having shaped their stones, get into the habit of asking: "*What's that*?" And they all answer in chorus to one another and each to himself: "*Chopper*!" In such a *collective cognizance*, the new semantic structure is confirmed for the whole group, in other words, the functional name of the instrument has become a *typical name* available to the group.

Finally, when this scene has been sufficiently repeated so as to be integrated in the enduring image of the group that each individual has within himself, this same scene motivates an *individual cognizance* by making the new sign available to the particular subject, so that the structure of the ideal typical image becomes a *psychological acquisition* which from now on will function normally in the very perception of the object and will thereby make possible the regulation of a *productive* activity.

We have analyzed earlier the internal dialogue which is also uttered out loud, where Piaget's child said to herself: "*What's that, Jacqueline, what's that*? . . . *A block*." We can now rediscover the origin of these very words in the inner dialogue of the prehominid ancestor at the very end of his development, in a moment of recollection, when contemplating the cut-stone instrument whose edge had almost attained a regular shape. This internal dialogue, also uttered out loud, was the *reflection*, by which the ancestor fixed by means of his consciousness of the *typical name*, the accumulated experience of hundreds of thousands of years of adaptive labor, and was thereby ready to reach the level of humanity by beginning the transition to the labor of production: "*What's that, Hunter, what's that*? . . . *Chopper*!"

III. THE ALVEOLUS OF THE DIALECTIC OF KNOWLEDGE

Introduction to Sentence Formation

The formation of the sentences as sentence, in the strict sense, which succeeds the syncretic stage just analyzed, raises an entirely new set of problems which demand to be treated within the framework of another analysis. We shall thus limit ourselves to some brief remarks.

With the cognizance of the typical name by which the differentiation of the substantive is achieved, the functional sentence, as an association of syncretic words, is transcended and we now come to the sentence in the strict sense of the term, based on the differentiated word. Thus in Gvozdev's list we have at 22 months: '*matsik kleça sidit*' ('baby chair sits' = baby is sitting in the chair), '*papa niska tsitatz*' ('daddy book reads' = daddy reads the book), etc. The child specifies a state of things in three words which was done in two at the preceding stage.

In fact, we have seen at the beginning of the preceding section, that the same subject at 21 months uttered elementary functional sentences with a somewhat comparable content although less explicit: '*baba kleça*' ('nanny chair', which we have translated approximately by 'nanny, that, she sat upon'), '*dada baba*' ('uncle water', or approximately: 'uncle, that, he is dripping'). It is obvious that at the same syncretic level, in the situation of the child sitting in the chair, we would simply have had '*matsik kleça*' ('baby chair') which more or less would have signified: 'baby, that he sat upon'. By the same token, seeing his father reading, he would have said in syncretic language: '*papa niska*' ('daddy book'), or approximately: 'daddy, that, he reads'. In other words, '*kleça*' ('chair') taken in the sense of object in motion would have sufficed to express the position in the chair and '*niska*' ('book') the act of reading. If now the child specifies: '*kleça sidit*' ('chair sits'), '*niska tsitatz*' ('book reading') this evidently means that '*kleça*' and '*niska*' can only be taken now in the sense of objects at rest and this is the reason why it becomes necessary to add '*sidit*' and '*tsitatz*'. In other words, the words '*kleça*' and '*niska*' are no longer syncretic words, but words differentiated as substantives, so that the sense of action which was presented in the image of the object in motion must now be transferred to '*sidit*' and '*tsitatz*'.

We can see that the differentiation of the substantive entails the appearance of the *sentence in the strict sense* with the relations of subject to verb, and of verb to complement, the complement being indistinctly direct or circumstantial as illustrated by the two examples just given. This new structure, however, is not yet completely constituted. In fact, as Gvozdev remarks, the verb comes only at the end, after the complement, which is contrary to the normal order of words in Russian. This order will only be acquired at about 23 months, which indicates a higher level. Thus Génia says at 23 months: "*dada nicot mouka*" ("uncle brings the flour"), "*itska litièla doundouk*" ("egg flown away trunk" = the egg has flown away behind the trunk). It is thus only now that with the regular word order that the *grammatical form* appears, and this is verified by the fact that it is also at this age that the differentiation of cases begins, notably for the accusative and the instrumental. Thus Génia says at 23 months: "*Daï kaskou*" ("Give the pap") while at 22 months he was still using the name with the undifferentiated ending in '*-a*': '*kaska*'. Also at 23 months we have "*baba pasla makom*" ("*baba pochla za molokom*" = nanny went to get the milk), while until then the milk had been indistinctly referred to by '*mama*'.

We can see that the first sentences in the strict sense, which develop immediately after the prehominid age, at about 21–22 months, do indeed imply in their general structure the *syntactical connection*, but this is not completely constituted since it lacks the *syntactical form* as grammatical form. In fact, the linguistic heritage of the prehominid stage comprises only the substantive as differentiated word, as typical name, so that the word which indicates action remains syncretic. Consequently, there cannot be any grammatical form in it yet, since the grammatical form can be established only between differentiated words. Thus in the example '*matsik kleça sidit*', it is easy to see that the child began by saying "*matsik kleça*" ("baby chair"). But since '*kleça*' has now been differentiated as a substantive, the child is obliged to add '*sidit*' (sits), and this is the reason why this word only comes at the end. We can therefore conclude that '*sidit*' is still a syncretic word taken in the sense of action and *functioning* as a verb and not as a verb in the strict sense. In fact, if it was a verb strictly speaking, that is, a differentiated word in the verbal form, it would have come before the complement '*kleça*', and not after it, since the verbal form in Russian implies precisely such a normal order of the parts of speech. It was, indeed, in that order that the word appeared in the sentences that the child heard around him, but he was unable to assimilate it, precisely because he understood that word as a syncretic word and not as a verb.

In short, the relations of subject to verb, and verb to complement analyzed earlier in Génia's sentences at 22 months were still only *functional, not formal*. We must thus consider these sentences as *sentences in formation* as opposed to *constituted sentences* which appear at about 23 months with the beginnings of the grammatical form, which is itself conditioned by verb differentiation.

The differentiation of the verb which is thus the end of the stage of sentence formation (21–22 months), is founded on the new progress of the developed indicative sign which appears at the very beginning of that stage following on the previous stage. Thus at about 20–21 months we can observe a new behavior in the child which consists of pushing a train of four blocks with the finger, imitating the observer, or himself laying out a train of two or three blocks in a line by pushing them (Brunet and Lézine). The regulation of such a behavior presupposes that the subject indicates to himself the form of alignment as the form, as of whole, of the motion to be imparted to the blocks, since any impulse from outside that form would only disperse them. We thus have here a new developed indicative sign which can be observed in the direction of the look as in that of the hand, and where the insistence in the first place of the formula bears on the moment of the form as the form of motion to be imparted to the object. We thus have the following meaning: 'the form of alignment (A) of the motion concerning this here', or AMT. We meet here for the first time the semantic structure defined by formula (6) presented at the beginning of this chapter: FMT, 'the form (F) of the motion concerning this here'.

If we return to phylogenesis, we may believe that a sign of this kind had to be constituted for the task of aligning stones for the construction of walls for shelters. In Lower Bed I of the Olduvai gorge, a circle of roughly piled stones was discovered which was probably used as a shelter by *Homo habilis* whose remains were discovered at the site MKI less than a mile away, at the same geological level.[71] A work of this kind presupposes that the subjects already possessed a sign to indicate to one another and each to himself the form of the acts (in their entirety) of stone alignment to raise a wall. It was probably a forward gesture of the hand, perhaps already with the pointed finger, which, by tendential projection, projects on the work place the line of the movements of construction to be accomplished. And it is this indicative sign of the alignment form (A), as form of the motion to be imparted to the object – AMT – which we rediscover in the gesture of the child who pushes the train of blocks with his finger. In fact, by this gesture the child indicates to himself the form of motion to be imparted to that train so that the blocks remain aligned, at the same time that he *produces* this movement of the blocks, since it is this very synthesis of the signifying act which projects the image of the form as the form of motion to be imparted to the object, with the realization of that form in the form of movement taken by the object itself, which defines the act of *production*.

Since the stone shelter of Olduvai just mentioned goes back to the beginnings of *Homo habilis*, site MKI being at the lowest level of the camp, we may believe that the construction technique of this kind appeared during the transition from the Kafuan to the Olduvian. This is confirmed by the fact that the sign which it implies is reactivated in the child at about 20–21 months, or during the transition from the functional sentence stage, which, as we saw, refers to the second phase of prehominid development, to the stage of sentence formation, which corresponds to the level of *Homo habilis*, 'man in the making'. Now the acquisition of such a technique presupposes much previous experience accumulated in the course of prehominid development. It is thus probable that the Australanthropi who evolved from the Kafuan could already more or less raise summary embankments.[72] And just as the cutting stone could be indicated by two different diffuse-syncretic words, according to whether it is considered as a whole as '*stone*' or as its useful part only as '*cutting*', so the embankment, which is a pile of stones confusedly aligned, involves two analogous words, either '*stones*' to designate it in its form as a whole, as a pile of stones and '*embankment*' to designate its useful aspect, in its alignment form. The meaning for '*stones*' established according to formula (2) is: 'this here in the form of a pile of stones (S) as it appears in its (presumed) motion', or TSM, and for '*embankment*': 'this here in the form of alignment (A) as it appears in its (presumed) motion', or TAM. The elaborative work of the embankment implies first of all, at the level of semi-shaping, a representative syncretic sign of the form of alignment as instrumental form, or according to formula (9): $T.\overline{TAM}$. And towards the end of the Kafuan the completion of shaping makes the determinate representative sign of the same form appear. It was probably a lateral motion of the hand executed with firmness, somewhat like the drawing of the 19 month old child, which clearly projects on the construction site where the stones are dispersed, 'the form of alignment (A) of this here (absent) as it appears in its (presumed) motion, represented with insistence on the object-stones (TSM)', which leads us to rediscover a variation of structure (20): $A\overline{TM}.\ TSM$. And, as in the shaping of the stone, the gesture is reinforced by the functional sentence: '*cutting stone*' (= 'in the cutting form, this stone!'), we have here '*embankment stones*', in other words: 'In alignment, these stones!' With the dialectic of comparisons which are developed in this sign, the embankment assumes a functional name that we can transcribe by '*wall*'. The 'wall' is defined here according to the structure of the first member of formula (22): $(T_1SM.\ \overline{AT_XM})$, which is stated: 'the object-stones (T_1SM) in its alignment form (A) relatively typical'. Finally, owing to the cognizance we have described, the word '*wall*' becomes a typical name according to structure (24): $(T_1SM.\ \overline{AT_XM})$, which is stated: 'the object-stones (T_1SM) in the ideal typical form of alignment'. We now find ourselves at the final stage of prehominid development. Actually, the almost regular wall which the last Australanthropus ancestor just elaborated, does not yet possess a typical, actually distinct form. But the cognizance of which we just spoke – "*What's that*? . . . *Wall*!" – which is a general representation of the alignment form $\overline{AT_XM}$, as realized on the object-stones (T_1SM), enables the subject to *ideally disregard* the actual irregularities, since the syncretic content (sm) of the image which reflects them has passed to the recessive state. Consequently, the subject indicates to himself, or he *perceives* the real wall *distinctly in its ideal typical form*. And it is precisely in such an *idealization*, obtained by the *work of language and consciousness* that the new model which motivates the transition to productive labor is constituted.

In fact, we have seen that the Australanthropi, who evolved from the last part of the Kafuan, had models for instrument elaboration at their disposal which consisted in representations of the insturmental form, where syncretism was being gradually transcended, but not actually eliminated. In other words, the model was being raised to a level of increasing precision, without being able, however, actually to stand out from the contingent traits of the natural shape of the raw material, since the typical image of the instrumental form, while more or less self constituted, remained united with the syncretic image of the material, resulting in the reappearance of a certain confusion. And since the contingent traits of the raw material appeared to each subject more or less differently depending on the syncretic image which reflected them, the resulting representational model which guided the shaping varied from one worker to another, so that the movements executed in collective labor by the various members of the group could not be standardized. With the appearance of the typical name and of the idealized image which it evoked, the workers for the first time had at their disposal a model which actually stood out from the contingent traits in question, in other words an ideally identical model for all, so that guided by this standard model their work procedures tended to assume a common appearance. In the case of the construction of the wall under consideration here, the workers, by piling up stones in one typical, ideally represented alignment pattern, tended to be in unison by each giving his gestures a form more or less parallel to those of his neighbor. Now, if in this unanimity, which soon is spontaneously established, one of them happens to deviate from the general pattern, the others call him back to conformity by a sign which is modeled on the pattern itself. In other words, we are no longer dealing here with the *lateral* gesture of the hand which ideally projects the typical alignment pattern of the wall under construction on to the construction site: $(T_1\,\text{sm.}\overline{AT_xM})$; *but rather* with a *forward* motion of the hand, modeled on the collective motion already more or less uniform in the alignment of the stones, which projects the insistent image of the alignment pattern no longer as the form of the wall to be constructed, but as the form of the very motion of its contruction, or *AMT*: 'the form of alignment (A) of the motion to be imparted to this here', or on the stones at the site. Naturally, when we speak of a forward hand gesture, the motion of 'forward' must be understood in an altogether relative manner, since, according to the position of the subject's own body who makes that sign, the modeling of the sign based on the collective work procedures of the ensemble of workers permits just as well a more or less oblique, almost lateral, gesture, as we can observe in its reactivation in the child pushing his train of blocks.

We may believe that at the same epoch of the transition to the Olduvian, the unity of the ideal model acquired with the typical name '*chopper*' (T_1sm. $\overline{CT_XM}$), also enabled the shaping motions of this instrument to take on a progressively uniform pattern, tending to give to the edge its typical distinct cutting form. As a result a new sign is produced which is modeled on that tendential figure itself, and which indicates to the clumsy workers the form of the cutting motions to be realized. We do not have any observations of child behavior which would enable us concretely to reconstitute that sign.

Its meaning can be defined by analogy with the one we have just established for the construction of the wall, or *CMT*, C now designating the cutting form insofar as the motions of stone cutting tend toward it, in other words 'the form tending toward cutting'. The meaning is stated thusly: 'the form tending toward the cutting motion (C) concerning this here', namely, the stone. Maybe it was a motion of the hand that successively imitated the symmetrical strokes of the stone-worker on both sides of the stone, which is altogether new in relation to the drawing gesture which we described for the preceding stage and which represented the cutting form simply by a straight or largely curved line.

The acquisition of the indicative sign of the form, as form of the motion concerning the object – FMT – finally causes the transition from elaborative work to *productive* labor. In fact, cognizance of this sign, which operates according to the dialectic of the three moments: sporadic, collective and individual, and which remains valid for the whole period of *hominization*, ends up by giving a new structure to the working act, insofar as it implies from now on its own *level* within itself as *consciousness of the form of the motion to be imparted to the object*. Thus it is indeed the image that the worker has projected from his head to the raw material, not only the image of the very form of the instrument to be obtained, but the image of the form of the motion of his shaping, which is realized in the material that is worked on. The laborer, says Marx " . . . realizes [*verwirklicht*] his own purpose in those materials. And this is a purpose he is conscious of, it determines the mode of his activity with the rigidity of a law, and he must subordinate his will to it."[73] It is, in fact, essential for the act of production, as a *technical* act, that consciousness does not limit itself to the representation of the purpose itself, but that it also specifies the *mode of activity* implied by it, for actually the purpose itself can be realized with precision only by means of the determination of that mode of activity. In fact, the form realized as *property of the produced object* is none other than the very form of the motion of productive labor, *insofar as this motion has been stabilized in that object*, and it is this same dialectic which defines the process of production in its most fundamental structure. "Labour" says Marx, "has become bound up in its object: labour has been objectified, the object has been worked on. What on the side of the worker appeared in the form of unrest [*Unruhe*] now appears, on the side of the product, in the form of being [*Sein*], as a fixed, immobile characteristic. The worker has spun, and the product is a spinning."[74]

The relation of movement to the object, which is rather obvious in the process of *human production*, a relation by which the property of the object is obtained, is only a *sedimented thing* as the *stabilized form of motion* is applied to it by the worker, so that the useful form of the produced object is nothing but the very form of the act of labor which is *objectified* in it. In reality this relation appears in the most general manner in nature itself, since all human productive labor is only a *conscious reproduction* of the spontaneous process by which natural phenomena are *themselves produced*. Every object being constituted by a multitude of particles in motion, its objective form, as "property at rest, in the form of being," can only be the general system of the motion of these particles taken at the present instant in its form of equilibrium as *motion provisionally stationary*. A property thus realized, of course, has only a temporary stability, since it is by virtue of its very nature, motion, and must necessarily transcend itself in the development of motion. This is precisely what Marx shows when he characterizes dialectic as the mode of thought which " . . . regards every historically developed form as being in a fluid state, in motion, and therefore grasps its transient aspect as well . . . ."[75] In other words, "every stable form" as a property of the object at rest is but a "transitory configuration" of the motion which is provisionally stationary. And the dialectical conception of the world thus defined very much conforms to "the dialectics of things themselves, of Nature itself,"[76] such as is demonstrated in the social practice of productive labor, where man *reproduces* in his own way things of nature by consciously repeating this fundamental process, where the motion concerning the object objectifies itself, in a more or less transitory way, by being stabilized in a configuration which is maintained for a while "as a property at rest, in the form of being." – "The worker has spun and the product is a spinning."

If "every made form" of the object, that is, the ensemble of its properties, is but the configuration of the motion implied in it, there is consequently nothing more to the object than its motion in various forms, so that knowledge of the object consists only in the knowledge of its motion. " . . . matter and motion *cannot* be known in any other way than by investigation of the separable material things and forms of motion, and by knowing these, we also *pro tanto* know matter and motion *as such*." [77]

It would be a serious mistake, however, to conclude from this that matter is itself reducible to motion so that there would be nothing left in the world except pure motion as such. We find ourselves right on the dividing line between materialism and idealism. In reality, motion is real only to the extent that it is motion *of matter*, and any attempt to posit it in itself, outside of matter, as pure motion, inevitably results in transforming it in an ideal motion of thought. In other words, motion implies a *subject* which moves, and this subject can only be matter, unless thought is substituted for it, in which case the whole of nature would become an attribute of thought. As Lenin says: "the mental elimination from 'nature' of matter as the 'subject' only implies the tacit admission into *philosophy* of *thought* as the 'subject' (i.e., as the primary, the starting-point, independent of matter)." [78] In Hegel, this abstraction of motion, which leads to its subjectification and mystification, comes to its ultimate end by being deified in the ecstasy of a pure "knowing" turning indefinitely upon itself as "knowledge of self" – which may as well be equivalent to a total immobility:

> Appearance is the arising and passing away that does not itself arise and pass away, but is 'in itself' [i.e. subsists intrinsically] and constitutes the actuality and the movement of the life of truth. The true is thus the Bacchanalian revel in which no member is not drunk; yet because each member collapses as soon as he drops out, the revel is just as much transparent and simple repose. [79]

In such a 'repose', motion is, in fact, completely suppressed, since it is not a more or less temporary rest in a transitory configuration of motion, but a *definitive repose* in the finished unity of the absolute Idea. And this is the reason why the Hegelian dialectic, after having exalted motion in consciousness, ends up in a pure conservatism on the plane of the real.

Thus if motion can be conceived only as motion *of matter*, unless one suppresses the self not only as real motion but as any kind of motion at all, it is then indeed matter which presents itself thereby as *subject*. Now, as subject of motion, matter must be defined independently of it. And since it has been shown that all of the properties which make up the content of matter, consist only of configurations of the motion implied in it, there remains one and only one determination which can define it as *matter* itself or *subject*, a 'property' so to speak in quotation marks, for it is absolutely original in relation to all the others: the determination of objective existence, the 'property' of *being an objective reality*, of existing outside of our consciousness. And we have seen in the introduction that it is this very determination of objective existence which Lenin demonstrated to be the *unique* property of matter recognized by philosophical materialism. This unique property is constantly presented by the indicative sign and is integrated in the very functioning of our sense organs, thanks to which matter is given to us in sensation as external reality independent of the subject: the '*this here*' of sense certainty.

In the development of the indicative sign, the subject follows the 'this here' in its motion which enables him to constitute a first image of matter in motion. And the various transformations of the fundamental formula of the developed indication largely reflect, at the very level of original syncretism, the dialectic of this motion of matter. If formula (1) TMF presents matter in a motion of a certain form, formula (2) TFM presents it in that very form as objectified into 'a property at rest' where the previous motion is temporarily stabilized into a determinate configuration 'in the form of being'. Formula (3) MFT presents the motion of matter in its transmission from the subject of the action to its object, and formula (4) MTF presents it as accomplished in the object itself. Finally, formula (5) FTM presents the objective form of the object as objectified form of the motion implied in it, and formula (6) FMT presents the form of motion in relation to the object. We can see that at the very beginning of knowledge, the developed indicative sign gives a confused but effective image of objective reality in its most fundamental structure, as *matter in motion* or *motion of matter*, "*motion whose every stable form is but a transitory configuration*."

Thus in the syncretic meaning of the developed indication ". . . we can (and must) disclose as in [an 'alveolus'] or a 'nucleus' ('cell') the germs of *all* the elements of dialectics . . ."[80] of knowledge as the ideal reproduction of the dialectic of things. The contradictions which it contains at the latent state will emerge progressively in the development of social practice, creating step by step new forms of language and consciousness, where the original semiotical structure is enriched each time with new *linguistic gestures* modeled on the material activity and the material relations of the workers, which, with the support of verbal symbolism, project a more and more encompassing and precise image of the external world.

We have seen the first stages of that dialectic at work in the prehominids, these 'transition beings' (Engels) between ape and man. In the first prehominid phase which we can place at the epoch of the transition from the Tertiary to the Quaternary, the developed indicative sign begins by being multiplied according to its first four formulas, in correlation with the development of *simple adaptive work*, namely, the employment of the natural and the prepared instrument which becomes a regular behavior thanks to the *freeing of the hand*. The quantitative development of this simple adaptive work in the still undifferentiated prehominid group ends in a first qualitative progress with the appearance of a vanguard, whose exchange of communications with the majority of the troop gives the indicative sign a new structure, as *representative indication of the 'this here' absent*. This original sign of representation, modeled on the newly arisen form of cooperation, provides a first solution of the fundamental contradiction found in the syncretic meaning of the developed indication between the moment of the object and the moment of motion, inasmuch as *the motion of the object ends up by making it disappear*. In fact, the representation of the 'this here' absent is developed by the constitution of the *representative sign of the absent object* – $T.\overline{TMF}$ (8) – completed by the *sign of syncretic representation of the instrumental form* – $T.\overline{TFM}$ (9) – and the *insistent sign of syncretic representation of the motion of the absent object* – $\overline{T}.MTF$ (11). And the ensemble of these first representative signs made it possible, at the syncretic level, to overcome the contradiction between the object and the motion in its most elementary form, namely the disappearance of the object outside of the actual field of perception. This first qualitative progress in the development of language and consciousness, which for the first time enabled the subject to ideally escape the narrow limits of the present given, thereby realizing the *freedom of the mind*, brought about the transition from the preparation of the instrument by direct manipulation of the raw material to its elaboration by means of a second instrument which opened up the second prehominid phase (Kafuan) which we can place in the first part of lower Pleistocene.

The use of the elaborated instrument develops the differentiation of special teams within the evolved prehominid group which gives to labor the form of a *complex work of adaptation*. The new contradictions which emerge thereby at the level of material activity and material relations among workers are reflected in the language by a series of misunderstandings in which the fundamental contradiction between the moment of the object and the moment of motion in the syncretic meaning of the developed indication is again developed: the word used by the speaker to indicate the motion of a certain object is understood by the hearer as representing the same motion of another object. Or moreover, the word used to indicate an object in a certain motion is understood by the hearer as representing it in another motion. Such a contradiction finds its solution in the formation of the *elementary functional sentence* as an association of two syncretic words, which in type I, specifies the relation between the motion and the object, and in type II, represents the motion between two objects either in the dynamic form of the action itself, or in the stabilized form of a spatial relation. Thus in the example given here: '*game rock*', the representation of the game behind the rock itself implied the image of a motion of going around the rock, since the spatial relation thus represented can be defined only by such a movement. And it is this image of going around the rock leading to the game, which resolved the contradiction between the moment of the object and the moment of the motion in the syncretic meaning of the two words used one after the other by the speaker in conformity with his own situation and understood by the hearer in an entirely different sense.

The development of the functional sentence in the form of enumeration and dynamic correlation ('*grasshopper jump hunter*') produced the representation of the absent biological need situation, which brings about the completion of the shaping of the instrument in leisure moments. In such a progress, the quantitative development of elaborative work passes to its qualitative transformation, since for the first time there results an actually distinct form of the elaborated instrument. And it is this qualitative progress at the level of material activity which is expressed in language by the formation of a new gestural structure: the *sign of the distinct representation of the form* – $T.\overline{FTM}$ (19) – whose development brings to light the second fundamental contradiction implied in the syncretic meaning of the developed indication, namely the contradiction between the moment of the object and the moment of the form, inasmuch as *the form as such remains constant, while the object appears in indefinitely varied aspects*. How, under these conditions, can a certain form *of the object* be defined? The problem was originally ignored since the instability of the syncretic image projected by the developed indicative sign enabled it to modify the moment of the form according to the configuration and the exigencies of the situation.

But with structures (19), (20) and (20′) ('*little, little Istine*') the moment of the form is distinguished on the first plane of the syncretic meaning and assumes thereby a constancy which it did not have in the primitive sign of the developed indication limited to the first four formulas. The image of the form thus determined enters in structure (21) where it makes possible the *comparison* between two similar objects: "*mist smoke papa*" as Piaget's child said, or "*stone cutting daddy*" in the similar sentence which we have reconstructed. The comparisons repeated with the mediation of a same determinate form (F) end up by representing it as common to an indefinite plurality of similar objects, in other words as a general form realized in the particular, which defines it as a more or less typical form in the meaning of the functional name: $(T_1 F_1 M. \overline{FT_X M})$. It is obvious that form (F), insofar as it is common to an indefinite plurality of objects, now assumes a fixity which immediately contrasts with the instability of the syncretic form (F_1). This opposition, which is first of all hidden by the rest of the confusion which still remains in the functional name, bursts out in equivocal situations where the concurrent representation of various relatively typical forms, also suggested by the sensori-motor data, requires a determinate choice, while no problem is raised for the syncretic form, which the subject, in such circumstances, simply indicates according to his mood at the time. The contradiction which appears here, only shows the fundamental contradiction, already present from the beginning in the meaning of the developed indication, between the moment of the object always moving and the moment of the form implying a certain constancy, but which did not yet come to light because of the instability of the syncretic image, where the moment of the form can always vary according to the needs of the situation. The contradiction now develops in the alternate, then circular, motion of various, more or less typical but competing forms in the equivocal situation in question, a movement which defines the questioning gesture, whose insertion in the already acquired schema of the functional name produces the first questioning structure: "*What's that*?" The dialectic of questions and answers transforms the functional name into the *typical name* $(T_1 f_1 m. \overline{FT_X M})$ which is a first solution to the contradition between the moment of the object and the moment of the form, by making the syncretic content $(f_1 m)$ with the contingent traits that it reflects, pass to the recessive state so that the typical form (F) is clearly distinguished as *form of the object itself*, independently of its accidental aspects tied to the situation. By virtue of such a new structure of language and consciousness, the elaborative work, which is now guided by an ideal typical image of the instrumental form, tends to assume a common aspect for all the workers. On this tendency is modeled a new linguistic gesture which indicates for the first time the form of the act of work itself, FMT, the cognizance of which produces the first *productive technique*, where the exact realization of the typical image of the instrumental form is secured by the conscious determination of the very movements employed to shape the material.

"The animal," says Marx, "is immediately one with its life activity. It does not distinguish itself from it. It is *its life activity*. Man makes his life activity itself the object of his will and of his consciousness. He has conscious life activity. It is not a determination with which he directly merges. Conscious life activity distinguishes man immediately from animal life activity. It is just because of this that he is a species-being."[81] With the acquisition and cognizance of the indicative sign of the form of the motion to be imparted to the object, the worker makes his very activity his work, in other words he "*makes his own vital activity itself the object of his will and of his consciousness*." Work has now definitely left " . . . that state in which human labour had not yet cast off its first instinctive form."[82] In other words, it has abandoned that spontaneous, *semi-animal* form which it still kept during prehominid development as adaptive labor in order to become a "*vital conscious activity*," the *labor of production*. And since it is this vital conscious activity which distinguishes man from the animal and it "is precisely through it that he exists as a species" we can say that, with the beginnings of the production of the instrument, verified by the already typical form of the cutter of the first Olduvian choppers, labor assumed "a form in which it is an exclusively human characteristic"[83]; the genus *Homo* has appeared. Production, however, still appears in an embryonic form, since it is limited to the useful part of the instrument. And if we refer to the morphological data, we do indeed see that in *Homo habilis* whose brain has clearly crossed the 'cerebral Rubicon' and whose hand remains "strangely non-human" (Napier), humanity takes only an embryonic form at the last stage of its gestation period: he is a child already, but only a child in the womb of his mother, 'man in the making' (*der werdende Mensch*).

In fact, the indicative sign of the form of movement to be imparted to the object – FMT (6) – by which labor as 'vital activity' of the laborer becomes 'the object of his will and of his consciousness' is still only a final variation of the developed indicative sign. In other words, syncretism is transcended in the representation of the object by the creation of the *typical name*; it is not yet transcended in the representation of motion, since the sign by which the subject indicates to himself the form of the motion to be accomplished remains a syncretic sign, even though syncretism begins to be transcended in it, since the moment of the form stands out in the first place of the formula. It is true that in the shaping of the instrument, the subject already has at his disposal the typical representation of the instrumental form, so that the indication of the form of the act of work, FMT, can be modeled on it. For the shaping of the chopper, for example, the indication of the form of the cutting strokes to be executed by the cutter is determined with some precision by the typical image of the form of the cutter to be obtained, already ideally represented in the stone. Such a determination, however, which achieves the typical form of motion only in an indirect manner, does not go beyond the narrow limits of the instrumental form, in other words *it does not go beyond the useful part of the instrument*. Consequently, the subject cannot yet come up with a total technique which applies to the whole stone. In short he has not yet raised himself to the level of *tool* production.

Tool production implies the shaping of the whole of the raw material according to a total typical form. Such a work presupposes that the subject represents to himself in advance a rather long series of well-determined movements. While the production of the Olduvian chopper only requires from 5 to 8 cutting strokes on both sides of the edge, the Chellean biface requires several dozen well-ordered strokes, and for each stroke the exact striking place, the direction and the force of the motion to be accomplished must be determined.[84] The subject must thus have in his mind a rather complex plan of action, for which the simple indication of the form of motion to be applied to the material, as we have described it, would be totally insufficient. Here the worker must be able to indicate to himself a series of operations of a determinate typical form which presupposes the *differentiation of the verb* as *typical verb*.

In short, the transition from the production of the instrument to the production of the tool involves the constitution of the sentence, which will be realized with the original dialectic of the production forces and the relations of production in the development of *Homo habilis* and his transition to *Homo faber*: 'full-fledged man', in Engels' words (*der fertige Mensch*). Only then, in the transition from the lower Pleistocene to the middle Pleistocene, is the gestation period of the genus *Homo* completed: in crossing the cerebral Rubicon, man leaves the maternal womb of nature in order to be born in a new world, the world of culture.

In order to mark this memorable birth in taxonomy, we propose to elevate *Homo habilis* to the rank of a sub-genus, so as to oppose him to the ensemble of all later forms. The genus *Homo* thus comprises two sub-genera: *habilis* and *faber*. Such a division enables us at the same time to get rid of the artificial opposition invented by philosophers between '*Homo faber*' and '*Homo sapiens*'. In fact, as sub-genus, *Homo faber* extends from Chellean man, *Homo faber primigenius* (Pithecanthropus) to man of the present type, *Homo faber sapiens*.

THIRD INVESTIGATION

MARXISM AND PSYCHOANALYSIS – THE ORIGINS OF THE OEDIPAL CRISIS

> " . . . these petrified conditions must be made to dance by singing to them their own melody." [Karl Marx, *Critique of Hegel's 'Philosophy of Right'*, Introduction, p. 134.]

Psychoanalysis is once again on the agenda since the events of May 1968 and it is, of course, up to Marxism to extract the 'rational kernel' of scientific observations that it contains, by freeing it of its illusions. The Marxists who have intervened in the discussion organized by *La Nouvelle Critique* have made a theoretical clarification and indicated the way to a solution. It is certain that Freud's work has been plagued by the ideological parasites of his time: psychological biologism and Durkheimian sociologism. Consequently, psychoanalysis finds itself incapable of understanding human life in its *real essence* as *social essence*, the true foundation of the individual psychism. Only historical materialism, by developing the theory of the socio-historical forms of individuality, can correctly interpret the rich material of objective data accumulated by psychoanalysis.

From the psychoanalytic side, the ideological tendencies of Freudian theory are categorically denied, a 'denial', which obviously follows a process rather well described by Freud, and would be better called a confirmation. André Green insists on the idea that it is not from an ideological bias, but because of the 'causality of the facts themselves," that one must start from biological references in order to define the "primitive psychological situations" which motivate the psychic realizations of individuals as well as the spiritual creations of societies. As a type of these primitive situations where, independently of the social relations of production, the 'dynamic source' common to the individual and social psychic life would be revealed, we cite the Oedipal structure which is presented as the universal structure, "a primary determinative, which is the relation to the begetters. Every individual, no matter who, is born with two parents, one of his own sex, the other of a different one. Here is a structure that nothing can change. You can vary the roles, the functions, the historical contexts, but within this triangle, you cannot vary the sexual relations, that is to say, the primary determinatives"

(André Green, pp. 24–25 and 27).[1] In other words the Oedipus complex, in spite of the variety of its manifestations, tied more or less obviously to the social forms of each epoch, would remain essentially independent of this historicity, since it rests on some sort of 'invariant' of biological origin, this 'primary determinative' which is the relation to the begetters.

This proposition seems obvious, even tautological, at least if one accepts the 'primary' character of the Oedipal triangle. The whole problem would be to know whether it does indeed have that character. If we refer to Freud's observations, the Oedipus complex appears at a rather late stage, since it does not begin until about age three in the little boy and a little later in the little girl. And it is preceded by a *pre-Oedipal* stage which appears at about age two, an age at which the object-relation is already clearly constituted,[2] but where the child's love for his mother does not yet imply any jealousy toward his father. Under these conditions, we can hardly see how it would be possible to consider the Oedipal triangle as a 'primary determinative'. Freud says precisely, at least with regard to the little girl, that "here the Oedipus complex has a long prehistory and is in some respects a secondary formation."[3] And after having analyzed various occurrences of the phenomenon, he concludes unreservedly: "In girls *the Oedipus complex is a secondary formation*."[4] Now, the case is evidently the same for the boy since he also goes through a pre-Oedipal stage, although it is shorter and less complicated. It was precisely with respect to the little boy that Freud recognized, not without some uncertainty:

> As regards the prehistory of the Oedipus complex in boys we are far from complete clarity. We know that that period includes an identification of an affectionate sort with the boy's father, an identification which is still free from any sense of rivalry in regard to his mother.[5]

We can thus generally conclude that for children of both sexes, the Oedipal structure is not an immediate given, but must be *constituted* as the result of a whole development. It is true that the object-relation to the begetters is always triangular, but originally that triangle was not Oedipal.

In *Totem and Taboo*, written in 1912, Freud had tried to explain the Oedipal structure, considered then to be original, by tracing it back to the anthropoid ancestor. The latter lived in families or groups of families, each dominated by a jealous male, who maintained his exclusive enjoyment of his females by force and eliminated his sons as they reached adulthood. These mores were probably prolonged in primitive humanity, to the time when the revolt of the frustrated sons ended in the murder of the father.

Then came the repentance of the Oedipal brothers who, in order to disavow their parricide, renounced the fruits of this crime by forbidding any sexual relation with the women who belonged to their father. Thus the incest taboo was established which marks the beginning of civilization .... Without insisting on the flagrant unlikelihood of this story, it will be sufficient to remark that it is refuted by the simple existence of the pre-Oedipal stage, discovered by Freud a decade later, the theory of which he always remained incapable of formulating. If human love begins, at the most tender age, by developing *without jealousy*, then we must obviously infer that primitive humanity was free from the beginning from the sexual antagonisms that divided the family of the anthropoid ancestor. The Oedipal conflict which emerges afterward thus could not be explained simplistically by the 'sexed relations', in the immediate and purely 'impulsive' sense, of the child to his begetters. It certainly presents an altogether different meaning, one which implies, on an actually *human*, hence *social* plane, a motion of mediations.

In reality, the very structure of the pre-Oedipal object-relation is absolutely incompatible with the fundamental concepts of Freudianism. Freud first of all conceives of love as love of self – autoeroticism and narcissism – and it is only by 'supporting' this original relation to self that the subject brings his desire to bear on the external object. The libido begins its formation in the baby at the breast, in his first autoerotic experiences, pleasure arising at the beginning as a 'supplement' or a 'gain' from the exercise of self-preservation functions, notably the function of feeding. This sum of libidinal energy thus accumulated within himself is first of all placed by the child on his own narcissistic image, and it is only on the basis of this self love, as investment of himself by himself that the object-relation is constituted as investment of the object. "Throughout the whole of life," says Freud, "the ego remains the great reservoir from which libidinal cathexes are sent out to objects and into which they are also once more withdrawn, just as an amoeba behaves with its pseudopodia."[6] As it is evidently constituted from that narcissism which itself is but the development of an original solipsism, the love of the object remains essentially egoistic; in the final analysis it appears only as a way of bringing back the object to himself, just as the pseudopodia of the amoeba end up by being reabsorbed with the object in the body of the animalcule. Under these conditions we understand very well how the child's first love, namely the love for his mother, gives proof of a ferocious exclusivisim which is particularly demonstrated in the jealousy shown toward his father. Within the framework of Freudian concepts, where psychological biologism gives an apparently scientific expression to the bourgeois conception of man and the world, the object-relation must necessarily be presented first in the Oedipal form. Now, from Freud's really scientific observations, it turns out in the end that this is not the case at all. The obvious impasse, to which the discovery of the pre-Oedipal relation has reduced psychoanalysis, shows that we are confronted here with a *crucial test*, which requires the transcendence of the *Freudian theory* in the name of the objective data of psychoanalytic observation.

I. THE ORIGIN OF THE PRE-OEDIPAL STAGE

If the pre-Oedipal stage plunged psychoanalysis into inextricable perplexities, we, on the contrary, find ourselves completely at ease in examining the problem from the positions of historical materialism. Engels has shown that it is totally impossible to represent the first human society based on the model of animal groupings torn by sexual conflicts. With the rather poor means available to humanity at its beginnings, the development of production required a close union and solidarity among the members of the group, incompatible with the structure of the anthropoid herd which is only a more or less provisional gathering of antagonistic families founded on the jealousy of the male.

The transition from the animal group to the first human society appears precisely as the resolution of the fundamental contradiction which tears apart the animal group, that is, the contradiction between the animal family and the herd:

> [In the higher vertebrates] the herd and the family are not complementary to one another but antagonistic. . . . The jealousy of the male, which both consolidates and isolates the family, sets the animal family in opposition to the herd. The jealousy of the males prevents the herd, the higher social form, from coming into existence, or weakens its cohesion, or breaks it up during the mating period; at best, it attests to its development. This alone is sufficient proof that animal families and primitive human society are incompatible . . .[7]

Because of such an incompatibility, the birth of human society was made possible only by the suppression of animal jealousy, in other words, the suppression of the animal family in favor of the herd. "For man's development beyond the level of the animals," continues Engels, "for the achievement of the greatest advance nature can show, something more was needed: the power of defense lacking to the individual had to be made good by the united strength and cooperation of the herd. To explain the transition to humanity from conditions such as those in which the anthropoid apes live today would be quite impossible ... Mutual toleration among the adult males, freedom from jealousy, was the first condition for the formation of those larger, permanent groups in which alone animals could become men."[8]

We may emphasize here the *social* or, in other words, *human* character of this still simplistic and immediate suppression of animal jealousy. If it is true that it ends up in some kind of "promiscuous sexual intercourse" which can only be deemed lacking in "prohibitions and restrictions," we must immediately add, as Engels specifies, that it must be understood in the sense that "the restrictions later established by custom did not yet exist," and by no means in the sense that "in everyday practice it necessarily implies general mixed mating."[9] The lack of sexual prohibitions would be precisely that unbridling of bloody rivalries which end in the constitution of the animal family, a more or less stabilized form of domination and hoarding of females by the strongest males. The communization of women in primitive humanity implies that animal jealousy, the foundation of the animal family, was overcome by the practice of collective labor. "The development of labour," says Engels, "necessarily helped to bring the members of society closer together by multiplying cases of mutual support, joint activity, and by making clear the advantage of this joint activity to each individual."[10]

It is precisely this still immediate and undifferentiated solidarity, as *first human social tie* based on the most elementary form of productive labor, which led to the establishment of the communization of women; for if this appears as "promiscuous sexual intercourse without restrictions and prohibitions" or "without shackles" when we compare it to higher forces of organization, it actually bears within itself a *fundamental restrictive rule*, namely the prohibition of sexual quarrels, which presupposes a collective discipline assuring the repression of all attempts to return to 'zoological individualism'. In the *Letter to Gorki* of November 1913, criticizing Gorki's view which made the idea of God the unifying element of human society, since it would have "repressed" "zoological individualism," Lenin states: "In reality, 'zoological individualism' was repressed not by the idea of God, it was repressed both by the primitive herd and the primitive community."[11] In other words, in the 'primitive herd', or 'herd of first men,'[12] sexual relations were no longer based on the strength and jealousy of the male as in the animal family, but precisely on general and mutual consent. Thus the particularly strong males who would want to possess a woman by violence and keep her against her will, as we commonly see in the apes, ran counter to the general indignation of the group and received a punishment from it which cured them of this *zoological individualism*.

This was evidently the *first decisive step in the transition from nature to culture*; the second to be realized, by the 'primitive community' or tribal society, was, as Lenin indicates, the prohibition of incest and the rule of exogamy.

It is easy to see that the pre-Oedipal stage in the child, characterized by love for his mother accompanied by a tender identification with the father which does not yet contain any sense of rivalry for the mother, refers precisely to this original form of primitive communism. The child begins by loving the persons of his environment in all simplicity, first his mother, then his father, because these persons reactivate in him the traces of the first social tie of human origins: the relation of immediate community [life] in the 'herd of the first men' or original community. The object-relation thus appears from the beginning as the internalized form of the human social relation.

The re-awakening of jealousy one year later with the Oedipus complex could thus refer simplistically to the animal rivalry of father and son in the anthropoid family arbitrarily transposed to the beginnings of human society. It is true that it is in the nature of the sexual conflict, at whatever level it be, to imply from a certain point of view a sort of falling back into animality. But since we already find ourselves here on the human plane, such a regression was itself made possible only because of the social contradictions which arise in the development of the original community, whose resolution will entail a new repression of 'zoological individualism', a repression which will be complete with the establishment of exogamy in tribal society.

II. THE GENESIS OF THE OEDIPAL CRISIS

Marxist authors, though holding many different points of view, nevertheless agree in placing the beginnings of the 'primitive herd' or 'herd of the first men' – what we have called the 'original community' – at the beginning of the lower Paleolithic, with the very birth of humanity in the form of *Homo faber primigenius* (*Pithecanthropus*).[13] This periodization coincides completely with our own conclusions drawn from the analysis of child language, according to which the stage of *Homo faber primigenius*, full-fledged man, corresponds with the child's fully constituted sentence at age 2, an age at which the pre-Oedipal object-relation occurs, which we just saw to be the reactivation of the original form of primitive communism. The contradictions which arise from the development of the original community, and which will lead to tribal society, had to be formed in the transition immediately following, or at the middle Paleolithic. In fact, if we refer to the archeological evidence, we find throughout the Mousterian (taken in the broad sense, from Riss to Würm) traces of a profound evolution of manners and customs: hut foundations of just a few square yeards,[14] which bear witness to the existence of households headed by couples, for such small lodgings could shelter only a single couple and its children. Thus from the lower to the middle Paleolithic, in *the very heart of the endogamic group, the transition from the communization of women to the pairing family was completed*.

The emergence of the pairing household obviously presupposes a form of division of labor between the sexes sufficiently developed to bring about the constitution of permanent couples within the original community. From the point of view of productive forces, the transition from the lower to the middle Paleolithic is characterized by the development of the use of fire and, for the stone industry, the progress from little splintered tools, which progressively replace the large multi-purpose biface. We may believe that the use of fire as well as the small specialized utensils, such as the scraper, gave rise to a 'household industry' for the preparation and preservation of foods, tanning hides, and the raising of children. Such an industry constitutes the material basis on which the association of conjugal couples or pairing families within the community is based. "In the old communistic household," says Engels, "which comprised many couples and their children, the task entrusted to the women of managing the household was as much a public, a socially necessary industry as the procuring of food by the men."[15] The pairing family thus appears as a *new social relation of production* which is constituted in the heart of the original community by virtue of the development of productive forces.

It is, of course, evident that the economic tasks created by the new division of labor no longer gave the two household partners the leisure for an unbridled sexual freedom as formerly. A certain limitation became necessary. We say a 'certain limitation' for the libertarian traditions of the original community always remained alive, the more so since the family economy was only beginning, and could only be inserted, in a rudimentary way, into the still predominant framework of the communal economy. We know, moreover, that with the so-called 'primitives' living now, there are many cases where the husband gladly loans his wife to a friend, or even to a stranger when he does not have to do so – not counting the orgiastic feasts where the sexual relations strictly forbidden in ordinary times become licit and sometimes obligatory. With the beginnings of the pairing family, during the Mousterian period, limitation of sexual relations to couples could only be relatively imposed. It nevertheless remained generally necessary, in order to enable the spouses to perform the economic tasks of household life. And it is quite clear that a restriction of this kind gave rise to all sorts of contradictions. *Jealousy*, which, in its strictly animal form, had been surmounted at the beginning of human society with the repression of zoological individualism and the establishment of the communization of women, is now reborn on a *social* basis, namely the contradictions which emerge with the new rule of the life of the couple in the pairing family.

As a first approximation, we can represent the family relations which appeared during the Mousterian period, by transposing, within the endogamic framework, certain elements of the matrimonial organization of the *Urabunna* in the nineteenth century in Australia.[16] Marriages in the Urabunna tribe occurred between cross-cousins, who were in relation to one another by virtue of their very birth, as virtual spouses or *nupa*. All the female cross-cousins of a man are his *nupa* or virtual spouses. And inversely, all the male cross-cousins of a woman are also her *nupa* or virtual spouses.[17] In fact each man has one or two *nupa* who are, strictly speaking, his spouses in the sense that they are especially given to him and live with him in his camp. Furthermore, he is also given a certain number of other *nupa* as secondary spouses or *piraungaru* to whom he has access under certain conditions. Finally, in case of need, he can have access to his remaining nupa, with the permission of their husbands, a permission which is generally given liberally. Inversely, each woman also has a special *nupa* as primary husband, and several secondary husbands or *piraungaru*.

In short, in the *Urabunna* tribe there is nothing similar to an exclusive right to sexual relations between a man and a woman. The couple's union implies only a *preferential right* which is on a par with secondary unions between *piraungaru*, and the occasional relations between simple *nupa*. In general, the men and women who are in the *piraungaru* relation live together by forming a particular group. It can happen, though rarely, that a primary husband seeks to prevent his wife's *piraungaru* from having access to her. But such an exclusiveness leads to a fight and the jealous husband is considered a clod.

The matrimonial system of the *Urabunna* tribe can be defined as a *group marriage with preferential unions*, and from the ethnological data, it is undoubtedly, so far as we know, the closest to the communization of women in the first human society. We thus propose to take as a starting point the structure of the *piraungaru* group, obviously excluding exogamy from consideration, in order to sketch an altogether provisional first 'model' of the social organization of these Mousterian villages, whose hut foundations we still possess.

We noted earlier, that, by virtue of their smallness, these habitations could not lodge more than one couple and its children. The endogamic community thus appears here as a complex of pairing families, where the partners live in a preferential union, at the same time that they have secondary or occasional relations with other members of the community. In such a situation, cases of jealousy can arise, since on the basis of the preferential right recognized for the relations of the conjugal couple there can arise a tendency to require an exclusive union. But these cases must have been rare, as in the *Urabunna* tribe, or even exceptional, since family life in this case has only just emerged from the communization of women.

In this first model of primitive society at the time of the origin of the pairing family, we have not yet seen what could have provoked that sharp crisis which is reactivated in the three year old child with the Oedipus complex. But we must now take into account a considerable factor which has left its imprint on the course of prehistory. Vallois's works have proven that throughout paleolithic times human life was shorter than it is now, and the life of women was even shorter than that of men. Nemilov had already drawn attention to this particularly premature mortality of women in anthropogenesis, which he called *The biological tragedy of the woman* (Leningrad, 1929).[18] The reason for this is that the acquisition of erect posture, with the development of bipedal gait, entailed a restructuring of the pelvis which, during the period of adaptation, considerably raised the rate of accidents during pregnancy and childbirth. Consequently paleolithic society suffered from a permanent disequilibrium, the number of women being far below that of men.

If we examine Vallois's latest statistics we see that in a collection of 29 Neanderthal men, in the broad sense, 18 of them were males and 11 females.[19] The sex ratio was thus 163, or 163 men for 100 women. This disequilibrium only slightly diminishes at the upper Paleolithic where we find 41 men for 29 women, or a sex ratio of 141. At the dawn of the Mesolithic period the sex ratio is still 126 (58 men for 46 women). And it is only during the Mesolithic period that the population equilibrium is re-established. The regular decrease of disproportion rates, from the Mousterian to the Mesolithic, makes these figures sufficiently trustworthy.[20] The biological tragedy of the woman with the demographic disequilibrium it created, evidently entailed major consequences for the organization of sexual life. While in the case of apes the number of females is regularly higher than that of males, which favored the formation of harems, in primitive society there were no longer enough women for the men. The problem was resolved at the beginning, as we saw, by the suppression of the animal family and the establishment of the communization of women. But from the time that progress in the development of productive forces in the heart of the first human society brought about the formation of a household industry, whence the birth of the pairing family, only part of the men could set themselves up in a household, so that they necessarily placed themselves in opposition to the others who were frustrated. In short, at the time of the transition from the communization of women to the pairing family, *the biological tragedy of the woman created the conditions of a real social tragedy*.

In Vallois's statistics, we find during the Mousterian period a woman more than forty years old and a man more than fifty.[21] These are, of course, exceptional cases, which should not be considered typical representatives of the population. In the collection of the upper Paleolithic, where longevity in general had nevertheless gone up, we do not find any woman who had lived beyond the age of forty. For the Mousterian period we can thus generally place the maximum age of women around thirty,[22] and for men around forty-five. Because of the insufficient number of women, we may admit that they were married at the nubile age. Consequently, the generation of married women was spread approximately over a period of slightly more than fifteen years, from the nubile age to thirty.

On the male side, it is probable that the young pubescents had to wait until their seniors had already established their households, before they themselves could be recognized as candidates. As a matter of fact, we know that among the primitive people who have survived to this day, the Australian tribes for example, the seniors have a definite advantage over the young, in matrimonial matters as well as in others, and we may believe that the same was true for Neanderthal men. Consequently, the generation of married men corresponding to that of women, such as we have defined it, and consequently spreading over a period of slightly more than fifteen years, would only begin at about age thirty and end at about age forty-five. In other words, *the young men had to wait almost fifteen years, from puberty to about age thirty, before they could marry*.

In the endogamic model drawn by analogy with the structure of the group of *piraungaru* in the *Urabunna* tribe, we saw that the couple's union, at the beginnings of the pairing family, was not exclusive but simply preferential, so that the married women normally had secondary or occasional sexual relations outside of the household. We may thus believe that such an arrangement enabled the young bachelors to more or less resolve their problem. Actually, at the end of the communization of women period, the union of the couple could accommodate extra-conjugal relations (of the women) only with already married men: for as each man already had his primary spouse, his relations with the wives of others were thereby sufficiently limited without taking into account the demands of reciprocity. But the case was altogether different for the young bachelors. At the beginning they undoubtedly had full freedom to participate in the extra-conjugal relations, but their ardor rapidly became a threat to the couple's union even if it were just a preferential one. From this we can conclude that the endogamic model proposed earlier was valid only for the circle of married people. In other words, in the Mousterian community, the pairing households maintained the unity between them with the system of preferential union combined with secondary or occasional relations. But this same unity made them agree to eliminate purely and simply the young bachelors who created disorder without the possibility of returning the favors granted them. Now, we have just seen that owing to the biological tragedy of the woman, men could not set up households until about thirty years of age. *It was thus the entire group of male youth which was totally frustrated*.

It is obvious that such a situation violently ran counter to the well-established habits of the original community where " . . . whole groups of men and whole groups of women mutually possess one another . . . "[23] It is true that the social harmony was preserved inside the group of married people, which shows that the opposition between the *ancient right* of the communization of women and the *new right* of the couple's union in the pairing family did not contain within itself any fundamental antagonism. But the demographic disequilibrium between the sexes could resolve the problem for mature men only, and this very solution exacerbated the discontent of their juniors, who, systematically condemned to abstinence, could not refrain from reacting sharply. In short, the biological tragedy of the woman gave to the opposition of two rights *the antagonistic form of a violent conflict of generations*.

At this dialectical moment of its development, the original community differentiates itself into three layers whose names are determined by function and social position. The mature, being married and having children, obviously assumed the name of '*Fathers*'. By the same token the reproduction function of women – who were all married at puberty – made them naturally assume the name of '*Mothers*', the more so since they were the spouses of the '*Fathers*'. The young men having the rights neither to the responsibilities and economic advantages of a household, nor even to sexual satisfaction, were thereby relegated to the position of minors, and consequently kept the name of '*Sons*'. As a result, the frustrated '*Sons*' projected their forbidden desires on the '*Mothers*' and thereby found themselves in a guilty rivalry with the '*Fathers*'.

The 'Oedipal' triangle which is sketched here on the basis of language does not, of course, have anything in common with the real relationship. In fact, the '*Mothers*' in question were between fourteen and thirty years old, in other words they belonged to the same generation as the young men called '*Sons*'. As for the '*Fathers*', so designated, they could not, in general, have engendered these '*Sons*' since they themselves were not married until they were nearly thirty years old, and hardly had the chance to go beyond forty-five, so that they could hardly have any children more than fifteen years old during their life-time Actually, under the conditions of the biological tragedy of the woman during the Mousterian period, *Oedipal incest was practically impossible*: the sons were almost always motherless before reaching the age of puberty. And in the social conditions just described, they were also generally fatherless.

The 'Oedipal' relation, just arisen in the endogamic community with the development of the pairing family, is thus essentially found only on the *semantic* plane where it expresses the contradiction of social relations sanctioned by language. However, if we represented by a diagram a typical model of the distortion of the matrimonial relations as they have just been described, we see immediately that if, from the point of view of age, the women who qualified as '*Mothers*' belong to the same generation as the young men called '*Sons*', it remains that, from the matrimonial point of view, they belong to the previous generation. (See Figure 1.)

In fact, it is the following female generation, that of the '*Daughters*' from zero to age fourteen which must furnish spouses to the '*Sons*' when they reach marrying age, or at twenty-nine. Now these '*Daughters*' come for the most part from the present generation of '*Mothers*' since these, married at age fourteen, could have had their first daughters at the age of fifteen. Consequently, these semantic '*Mothers*' are in the position of *virtual mothers-in-law* in relation to the '*Sons*' in question. We say 'virtual mothers-in-law' rather than 'future mothers-in-law' for in general they will already be dead, when these '*Sons*' marry their daughters. This virtual relationship, however, is nonetheless effective since it is established in reality. And since on the level of Neanderthal language the meanings of 'mother' and 'mother-in-law' are necessarily confused, we may believe that this effective

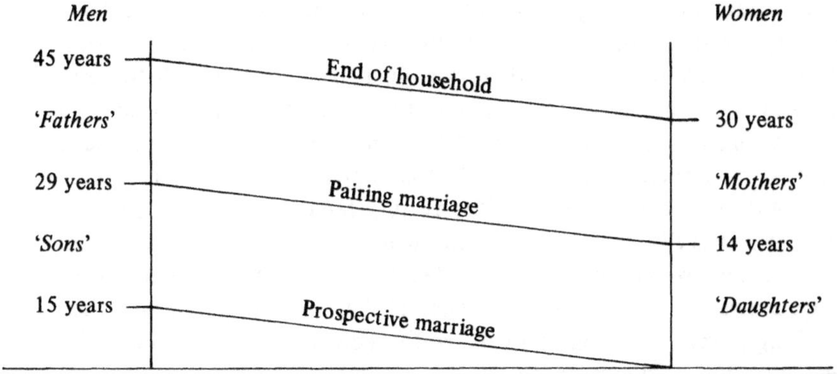

Fig. 1. Typical model of the distortion of matrimonial relations during the Mousterian period

relationship to the virtual mother-in-law gives some consistence to the appellation '*Mother*' given to the young women who by their age belong to the same generation as the '*Sons*'.

Moreover, there is a case where the virtual mother-in-law becomes a real mother-in-law. If a woman who has had her first daughter at age fifteen lives long enough to reach age twenty-nine, she will actually witness the marriage of that daughter, who just turned fourteen, with a '*Son*' of twenty-nine. We see that in this case, *the mother-in-law and the son-in-law have exactly the same age*.

There still existed in the nineteenth century in the Australian Arunta tribe a very unusual custom according to which every woman had to be engaged to a man as *Tualcha-mura*, a word which means actual or future mother-in-law. The engagement generally occurs in childhood. When two families have one a boy and the other a girl of approximately the same age, and have agreed to the engagement, they proceed to perform a ceremony in which the two children are promised to each other not, as might be expected, as future spouses, but as future son-in-law and mother-in-law *Tualcha-mura*. Consequently, the boy must wait until his *Tualcha-mura* is married, has her first daughter, who then becomes nubile, so that he can then marry this daughter.[24] We can see that according to this custom men marry rather late with women considerably younger than they are.

If we take the case where the *Tualcha-mura* is precisely the same age as her future son-in-law, and if we assume that she has her first daughter at age fifteen, as a result, she and her son-in-law will be twenty-nine when celebrating the marriage promised in their childhood. We thus have here exactly the same concrete situation that we reconstructed for the Mousterian period, by taking the case where a woman lives long enough to be able to witness the marriage of her first daughter. This apparently bizarre custom of the engagement with the *Tualcha-mura* can thus be understood as the survival of a state of affairs which was imposed by the force of events at the time of the formation of the pairing family in the conditions of the biological tragedy of the woman and which was sanctioned by tradition. In fact, in the Arunta tribe of the nineteenth century, the demographic equilibrium between the sexes had been re-established for a long time, and there was no longer any good reason to oblige the young men to wait regularly until age thirty before they married.

It should be noticed that in the Australian tribes every man must have the greatest respect and the greatest reserve towards his actual or virtual mother-in-law. In general he does not even have the right to speak to her. The extraordinary severity of this taboo suggests that it was intended to repress particularly violent temptations in regard to the mothers-in-law. Now, if in the Arunta tribe, as we have just recalled, the *Tualcha-mura* is approximately of the same age as her son-in-law, so that her seduction might be feared; the same is not true of other tribes, the *Urabunna* for example, where the mother-in-law necessarily belongs to the previous generation since she is the father's oldest sister. The taboo can therefore refer only to a distant past. As a matter of fact, during the Paleolithic, since the young men were obliged to wait for almost fifteen years before they could marry the daughters of the women of their own generation, they could not resist lusting for their future or virtual mothers-in-law during this long abstinence.

III. THE BIOLOGICAL TRAGEDY OF WOMAN AND THE BIRTH OF *HOMO FABER*

We must now retrace our steps for a closer examination of the biological tragedy of the woman whose considerable importance in the beginnings of human society is becoming apparent.

As we recalled earlier, all authors agree in attributing the premature mortality of women in prehistory to the accidents of pregnancy and childbirth consequent to the restructuring of the pelvis in the transition to erect posture and the development of bipedal gait. However, in order to specify the influence of the biological phenomenon on social evolution, it would be necessary to determine exactly when it appeared.

The idea has sometimes been advanced that the origins of this tragedy are identical with the acquisition of bipedal gait itself, so that it would have begun with the Australanthropi, since it was precisely they who introduced the hominid family through the transition to erect posture. The adaptation to bipedal gait, however, appears as a complex process involving several stages. This can be observed in children who learn to walk, from the second to the third year, and the study of fossil remains entirely confirms this testimony of ontogenesis. That the Australanthropi held themselves erect is beyond any doubt whatsoever, but the arch of the foot, still not pronounced, shows that their gait was not fully developed. Consequently, the process of restructuring the pelvis could only have just started and this does not necessarily coincide with the beginning of the biological tragedy of the woman.

In fact, if we examine the hip bone of Australanthropus, we notice a remarkably unequal evolution between its upper part and its lower part.[25] While the upper part, or *ilium* is considerably shorter and broader in relation to the ape – which does show that it supported the weight of the trunk, which consequently was normally in the vertical position – the lower part, or *ischium* is only slightly shorter in relation to that of the anthropoid and is much longer than that of man. In other words, Australanthropus already had a human pelvis in the upper part, but the lower part remained simian. Now, it is quite obvious that the difficulties of parturition were connected with the restructuring of the lower part of the pelvis and not of its upper part. And since this lower part had not yet been noticeably transformed, we can infer that in the Australanthropi, the biological tragedy of the woman had not yet actually begun. It is true that the transition to bipedal gait was already sufficient to create difficulties for carrying the fetus during the gestation period, which probably increased the number of miscarriages. But such accidents were rarely fatal. As a whole, it undoubtedly could result in a certain diminution of the female population. But in all likelihood, it could not yet result in anything particularly tragic.

The restructuring of the lower pelvis, or little pelvis, must, on the contrary, have entailed altogether disastrous consequences. From Australanthropus to man, the *ischium*, or lower part of the hip bone, was considerably shortened. There resulted an important contraction of the lower strait of the maternal pelvis, in other words a real strangulation of the birth canal. At the same time, the development of labor and language created new cortical layers which increased the volume of the brain. In other words, as the exit of the maternal womb became narrower, the fetus at birth presented a larger head. There obviously followed a *widespread dystocia*, which frequently entailed the death of the mother and child, when the birth occurred at full term. Only *premature fetuses could come out in a relatively easy manner.*

We see that the fetation of the human child, which has certainly played a primary role in *hominization*, consisted in a *natural selection resulting directly from the biological tragedy of the woman*. Adaptation consisted in the acquisition by the female reproductive system of a new functioning rhythm which enabled it to give birth to the fetus normally *before full term*. As we have seen, this adaptation took considerable time since the biological tragedy of the woman ended only in the Mesolithic. During this whole period, female mortality in childbirth, or the proportion of blocked childbirths, diminished more or less regularly, as we can infer from the decreasing curve of the sex ratio from the Mousterian to the Mesolithic. We can reasonably extrapolate this curve to the beginning of this process, as a result of which biological tragedy of the woman and the fetation of the child had to be most intense *at their first stages*. Now, we know that fetation has largely favored *hominization*, and the growth of the brain in particular. It is logical to believe that its first stage, which was the most intense, had to be precisely placed at the most intense moment of the cerebral development, when the famous crossing of the *cerebral Rubicon* occurred, where humanity, still in the gestation period of *Homo habilis*, effectively originated with the formation of *Homo faber primigenius* (*Pithecanthropus*). The volume of the brain leaped ahead with an increase of about 200 cm^3, suddenly increasing from less than 700 cm^3 to more than 800 cm^3. If we place the beginning of fetation in this period, it follows that *the appearance of the biological tragedy of the woman presided over the very birth of genus* Homo.

The shortening of the *ischium*, which was the immediate cause of this process, evidently could not be explained by a simple contingent mutation, but results precisely from the development of social labor during the previous period. In effect, because of the progress of the hunt brought about by the elaboration, then the production of the instrument, Australanthropus, and after him *Homo habilis* had to cover greater and greater distances each day in search of game. Now, the gait of Australanthropus, hindered by the excessive length of the *ischium* inherited from the anthropoid ancestor, was not effective, involving big expenditures of energy for a rather short distance. The adaptation to the growing requirements of displacement brought about the shortening of this bone, which made possible the acquisition of the proper human gait.

We know that the human gait contains essentially three phases. In the first, one foot begins a stride by moving forward while the second serves as a basis of support. In the second phase, while the first foot continues its stride, the second foot, or basis of support, lifts itself on the toe, at the same time pressing or pushing itself backward [and downward] which projects the body forward. Finally, in the third phase, the first foot completes its stride by touching the ground with the heel, and the second leaves the ground by the toe in order to move ahead in its turn. This rhythm, which assures the human gait of its suppleness and efficacy, gives it the definition of *walking by strides*. Each stride implies essentially that the weight of the body exerts its pressure on the ground by progressively passing *from the heel to the toe* of the supporting foot, the latter *thrusting backwards* so as to project the body forward. This is expressed in the composition of the English phrase to designate the human gait: "heel-and-toe."

Now the thrusting backwards of the supporting foot, which is an essential part of this whole motion, was not possible for Australanthropus because of the length of the *ischium*, which stops the backward extension of the thigh. Consequently, Australanthropus very probably walked *flat-footed*, a gait which we can more or less find again in the child who takes his first steps between 14 and 20 months *at the prehominid age*: the supporting foot remains motionless by resting on the whole sole, then leaves the ground all at once without pushing on the toe. Consequently, the foot which moved forward cannot make a stride, since the stride implies the body's forward projection, which is accomplished precisely by the backward thrust of the supporting foot with the help of the toe in back. In other words, the child at prehominid age walks with *flat-footed, little steps*; and Australanthropus, undoubtedly walked in the same way which, with a fast rhythm, produces a precipitous and jerky gait causing great tiredness over a short distance.

We may believe that the extension of the hunt following the elaboration of the instrument among the Australanthropi, evolved from the Kafuan, brought about a development of the gait which has progressively made the foot and the leg more supple. And as a result of prehominid development, this suppleness has made possible a first progress in the development of bipedal gait which we can try to represent by analogy with the child's gait after 20 months, that is, at the age of *Homo habilis* or the 'habilian' age. The child lifts the support foot on the toe, by more or less pushing backwards on the ground but *without clearly thrusting the foot backwards*. Such a walk is already less stiff and less tiring than the flat-footed gait, but it obviously does not yet have the efficacy of a walk by strides, since the body is not yet projected forward. We may assume that it was practiced by *Homo habilis*, an hypothesis which would be confirmed by the structure of the terminal [ungual] phalanx of a big toe discovered in the Olduvai Gorge and most probably belonging to *Homo habilis.*[26] In fact, the incline and the torsion of the head of this bone in relation to its shaft, characteristics which rank among those properly belonging to man, are indications of the backward push exerted by the toe on the ground, just at the moment when that foot is about to move forward. It is very probable, however, that *Homo habilis* could not yet readily thrust his leg backward, since such a motion would presuppose as already acquired the maximum shortening of the *ischium*, which, as we saw earlier, could only have occurred at the moment of the transition from *Homo habilis* to *Homo faber primigenius. Homo habilis* probably had a somewhat shorter *ischium* than the Australanthropi, but not yet to the point of enabling walking by strides. However, they had the necessary condition to acquire it, since they already lifted the heel by pushing backward with the toe, a motion which, in a fast walk, tends to thrust the foot backwards more and more. As this tendency was impeded by the still excessive length of the *ischium*, its development resulting from the development of the walk required by the progress of the hunt in *Homo habilis*, it brought about the maximum shortening of that bone, which made the transition to walking by strides possible, but at the same time reduced the lower strait [outlet] of the maternal pelvis to such dimensions that the full-term fetuses could no longer come out normally. In short, the restructuring of the little pelvis [pelvic outlet], progressively imposed by the long evolution of bipedal gait during the epoch of prehominid and *Homo habilis*, reached the nodal point where quantitative development took the form of a qualitative leap, exploding the biological tragedy of the woman, whence the fetation of the child, which enabled, with the crossing of the *cerebral Rubicon*, the birth of *Homo faber*.

Naturally, in this *historical dialectic*, as the *bio-social* dialectic of hominization, we must carefully distinguish [on the one hand] between the *logical dialectic* of fundamental movement, that is the development of labor and of the social relations with language and consciousness, and on the other hand the *purely historical determination*, that is the biological tragedy of the woman and the fetation of the child which precipitated the movement in the manner of a *catalyst*. It is clear that the soaring cerebral development of *Homo faber primigenius* came essentially not from fetation, but rather from the development of instrumental activity, which, at the end of the stage of *Homo habilis*, tended to reach a qualitatively superior form with the transition from the production of the instrument to the production of the tool. The fetation of the child, which itself appears through the mediation of the biological tragedy of the woman, as the *biological echo of the social development* of bipedal gait, only accelerated the evolution of the brain which was already determined essentially by the development of labor and language. However, this *catalytic action* of the biological event, as purely historical determination, nonetheless played a major and somehow *decisive* role, in the sense that it gave to the logical dialectic of development its *concrete form of realization in history*. And this is true not only for the formation of the human brain, but also for the constitution of the first human society.

In fact, owing to fetation, man's young come into the world in a totally helpless state, which leads to the establishment of entirely new relationships between mother and child. While the little ape clings to his mother, and after a little while is able to move around by himself, the human baby must be held in the arms and carried for years. The result is a general encumbrance for all the women, for, because of the high female mortality rate, many of the mothers died leaving children in early infancy in charge of others. In other words, at the final phase of the evolution of *Homo habilis*, when the tragedy of childbirth became acute, *the women, all having children, could no longer follow the men in hunting expeditions*, especially since these expeditions got longer and longer. The women therefore remained with the young, gathering food around the camp. And since such a group was incapable of defending itself against wild animals, some of the men had to be left with the group in order to protect it, while the most vigorous hunters went afar in search of game. In short, for the first time in the herd of *Homo habilis*, which undoubtedly until then had for the most part kept the structure inherited from the animal period, there was formed a *division of labor with the temporary separation of two groups*. Now, the animal family, which is the basic nucleus of the animal group, required constant vigilance on the part of the male, master of the family, since it was founded only on jealousy. Obviously it could not withstand the daily separations, and this was the historical condition which decisively favored the transition to the first human society.

By relying on more or less variable estimates from different authors we can assume, for example, a herd of about a dozen mature adults (counting from puberty). The sex ratio was at its maximum, since we are at the threshold of the biological tragedy of the woman: we can, without exaggeration, estimate it to be about 200, since during the Mousterian, after more than half a million years of adaptation, it was still at 163. For the population in question, we would thus have eight men for four women. And as for the children, we know that in primitive times their number far exceeded that of the adults. According to the figures given by Vallois, the site of Sinanthropos at Choukoutian gave 15 children for seven adults or adolescents. The Ibero-Marusian site of Taforalt (Morocco) had 93 children for 68 adults, but it seems probable that the real proportion must have been greater, for the bones of children crumble easily and are less susceptible to fossilization. In the pygmies of the Congo and the Cameroon, who have kept their primitive mode of life, the hunting groups generally comprise about a dozen adults with twice the number of children.[27] In the present case, we can thus hypothetically represent, with some likelihood, a herd of eight men, four women and 20 children. We may suppose that at the moment of daily separation, the adults divided themselves equally, or six for the hunt and six for the gathering of food. In other words, two men remained to protect a group of 24 women and children.

Since the heads of families – we shall call them the 'masters' since they dominated the others – were by definition the strongest individuals of the herd, they obviously headed the hunting expedition, so that the two men left with the food-gathering group were necessarily chosen from among the bachelors. In the presence of the masters, these could not come close to their women, but it goes without saying that some of the bachelors, who went with the women to look for food, had full access to them.

Two cases may present themselves. According to the ancient practice of zoological individualism, finding themselves face to face with four women, the two bachelors had normally to compete with each other in order to get the greatest possible number of women. However, such quarrels necessarily included the difficult task entrusted to them, which was that they both had to guarantee the security of 24 practically defenseless individuals. Now, after the millions of years of the development of language and consciousness in the activities of labor and social relations, man in the making had acquired habits of union and solidarity which, under certain conditions, could prevail over animal jealousy. We may believe that in the difficult situation of the protection of the food-gathering group, and having more than a sufficient number of women, the two bachelors made social reasons prevail over animal instinct, renouncing sexual rivalries in order to seriously perform their social obligations.

The communization of women, appearing thus under particularly favorable conditions, becomes a habit for the group of bachelors, for we may believe that they took turns in guarding the food-gathering group, since this task brought them sexual satisfactions of which they were once again deprived when the hunters returned. Such an agreement is sanctioned in the language by the express interdiction, proferred by the bachelors among themselves, of any return to zoological individualism. Only the masters, or heads of families, remained outside of the movement, since, being the most vigorous, they continued to specialize in hunting, and again took exclusive possession of their women when returning to camp. But the bachelors agreed among themselves to definitively impose the communal law. And since the women found their liberation in this law also, they all formed a general coalition, which, by the strength of the union, repressed the zoological individualism of the masters.

We have just expounded the favorable circumstances which may have accelerated the transition from the last group of a still animal form, the herd of *Homo habilis*, to the first, strictly human community, at the beginning of the Chellean. As we have already indicated, however, the fundamental dialectic of the movement can be understood only in terms of the progress in the production of the instrument, which, at the end of the evolution of *Homo habilis*, had given rise to the first elements of tool production, whose development was incompatible with the zoological matrix in which the embryo of human society had matured until then. In fact, tool production implies the conformity of the worker's gestures to a *plan* that is identical for all the members of the group, which was not feasible with the characteristic divisions and conflicts of the animal group. The quarrels about the apportionment of food and women had, undoubtedly, been attenuated at the prehominid epoch and particularly with *Homo habilis* because of the development of collective labor, syncretic language and group consciousness. Such an attenuation was sufficient for the elaboration of the instrument, followed by its production, which still require only conformity of the act of labor to the simple image of the instrumental form, as the form of the *useful part* of the instrument, or of the procedures to obtain it. But tool production, the biface being the first example, implies a *standard plan* representing invariably the succession of numerous operations necessary to give a determinate form to the material *as a whole*. The first elaboration of such a complex representation was made possible only by a unanimous and permanent social agreement, which could be established only within the communal structure, as the original form of human social relation, sanctioned by the repression of any return to zoological individualism.

The foundation of the first human society with the rule of the total communization of women and goods thus appears as the definitive and radical solution of the fundamental contradiction of the whole gestation period of the genus *Homo*, namely: the contradiction between the *social character of labor* from its beginnings and the *zoological individualism* which persisted in the *appropriation relations* still dominated by the hoarding habit inherited from the anthropoid ancestor. And since such a solution could intervene only when the instrumental forces, fully developed in the production of the instrument as the result of the development of *Homo habilis*, elaborated the first elements of the *production forces of the tool*, which required the elimination of zoological individualism, we may view the whole of this process as the inauguration of the fundamental law of every human society: *the law of necessary correspondence between the productive forces and the relations of production*. It is obviously this original dialectic, which takes into account the emergence of the human – the biological tragedy of the woman and the prematuration or fetation of the child having played only a role of facilitation – which precipitated social development. The establishment of communal relations realized the *first liberation of productive labor*, which made possible the decisive transition from the production of the instrument to the production of the tool.

Since the first tool was the biface, a characteristic piece of the Chellean period, we shall try to follow, on the basis of the proto-biface's embryonic form at the time of *Homo habilis*, the articulations of the dialectic whose general aspect we have just outlined. At Olduvai Gorge, the proto-biface, absent in the lower part of Bed I, becomes more and more numerous at the upper levels and at the base of Bed II.[28] Since the base of Bed II indicates the completion of the period of *Homo habilis*, we can consider the proto-biface as the result of the *first phase of the development of Homo habilis*, which became the specific product of the *second phase*. Here we are dealing with a highly evolved instrument, which, as a biface, possesses two cutting edges joining in a point, but which is distinguished by the fact that the part held in the hand, in other words the handle, still retains the natural form of the stone. In the biface, on the contrary, the handle is also shaped by cutting. It is evident that the cutting of the handle makes possible the lengthening of the double cutting edge by giving the point a sharp form, while in the proto-biface the edges are short and the point is not distinct. In other words, the proto-biface still results only from a differentiation of the chopper whose unique edge took a convex shape that made it appear as an obtuse angle. Such a differentiation, which arose toward the middle of the development of *Homo habilis*, remains within the limit of the structure of the instrument, since it still leaves on the material an unworked surface nearly equal to the surface of the double edge obtained by cutting. The transition from this last form of the instrument, the proto-biface, to the most primitive form of the tool, the biface, implies some sort of fiddling with the instrument which lengthens the double cutting edges of the handle, so that the two edges tend to join each other, thus forming a sharp angle at the tip.

It is clear that such fiddling, which obviously increases the effectiveness of the instrument, required an effort of sustained attention and a perseverance in the work which was realized with difficulty in the state of overt or covert general rivalry. This state of affairs was always possible in the herd of *Homo habilis* owing to the persistence of zoological individualism, which distracted the workers from their productive labor. The perfection of the proto-biface which brings about the transition to the biface, will therefore be fully achieved only after the elimination of zoological individualism by the establishment of the communal structure which alone could fully guarantee the peace and social unity necessary for decisive progress in production. We must notice, however, that at the very level of the herd of *Homo habilis* rivalry did not appear uniformly among all individuals. If the masters, or heads of families, had to keep a constant watch over their women in order to prevent a possible infidelity, the bachelors, in return, were by definition indifferent to such worries. It is true that whenever the situation was conducive to it, they tried to seduce the masters' women, just as they competed for the apportioning of women who had become available. But outside of these rather rare cases they were freed from sexual rivalries, and consequently, at the moments when the herd was at rest, could spend more time than the masters on the production of instruments. We may thus believe that it was precisely the bachelors who, toward the end of the development of *Homo habilis*, began fiddling with the proto-biface which led to the biface. Consequently, *the group of bachelors now represent the first elements of the productive forces of the tool which germinated in the heart of the herd of* Homo habilis *at the end of its evolution, and whose development was incompatible with the survival of zoological individualism*.

In the group of anthropoid ancestors, the bachelors were obviously the oppressed, par excellence, since they were brutally deprived of sexual satisfaction, bullied in the apportionment of food and even sometimes purely and simply expelled. Their situation had undoubtedly improved in the prehominid and *Homo habilis* group because of the attenuation of zoological individualism, which we indicated earlier. We can also note that the regular use of the instrument enabled them to defend themselves, which was not possible at the anthropoid level. When two apes fight, victory is practically won in advance by the stronger. This was no longer the case with the Australanthropi, since the use of weapons enabled the weaker to sometimes prevail over the stronger by skill and cunning: *David killed Goliath by throwing a stone at him*. Of course, this was not generally the case, but the zoological domination of the strongest began to be contested, which undoubtedly imposed some concessions on them. It is nevertheless true that, essentially, the weakest continued to be frustrated, particularly in sexual matters. We may thus believe that the struggle between oppressed and oppressors was to continue during the whole period of prehominid and *Homo habilis*. It actually led to a dead-end, however, for in the case where a bachelor won a victory over a master, he simply took his place, which just perpetuated the old order of things.

But when, towards the end of the development of *Homo habilis*, the bachelors began to use their leisure time to fiddle with their instruments, especially the proto-biface, and thus became the representatives of the new productive forces of the tool, their struggle against the masters took on an authentically *revolutionary* sense, since it now tended to *liberate productive labor from the zoological law of the strongest, by the establishment of new, really social, relations of production*. And we have seen that, favored by the new situation created by the biological tragedy of the woman and premature childbirth, they elaborated the new communal rule which was to prevail over the zoological domination of the masters, thanks to the cooperation of the equally oppressed women.

Thus, if, according to Engels, *labor* and *language* were "the two most essential stimuli under the influence of which the brain of the ape gradually changed into that of man",[29] there had to be a true *revolution* in order to give birth to the first human society which completed its gestation in the womb of the still animal herd of *Homo habilis*. It was through a long hard struggle that the bachelors, creators of the first rough-hewn tools, brought to an end the old zoological past, by eliminating the power of the hoarding masters with the help of the *Homo habilis* women who also craved liberty. We may now give that revolution the name '*faberian revolution*' since it opens the way to *Homo faber*.

The establishment of the original communism undoubtedly required numerous and repeated efforts, for it was impossible to suppress all at once the ancestral habits of recent animality. When a generation of bachelors came to compensate for their individual weakness by uniting with one another in order to impose the new relations, the problem became dormant in the following generation, where the strongest individuals were naturally tempted to repudiate the communal law in order to re-establish the privileges of physical strength. The definitive victory of the revolution was undoubtedly acquired only during a period of want, when difficult living conditions obliged the evolving community to impose a strict collective discipline under penalty of extinction.

At Olduvai Gorge, above the base of Bed II which contains the last deposits of *Homo habilis* of the end of the lower Pleistocene period, there is an Aeolian bedrock which testifies to a drought period indicating the transition to the middle Pleistocene. Above it follows the Chellean deposits from the middle Pleistocene. It was probably during the difficult times of this drought that the transition to human society was finally accomplished. As a matter of fact, in the face of the harsh trials imposed on the evolving communities, any resurgence of zoological individualism became a serious threat to collective life itself. The communities which let the internal quarrels come up again thereby condemned themselves to dispersion and annihilation. The only communities to survive were those which could consolidate the revolutionary conquests and definitively establish this immediate form of *equality*, which guarantees the salvation of the humanity just born, so to speak, and which will dominate the whole course of prehistory, and whose memory will remain afterwards in class society as a nostalgic aspiration of the popular masses.

At the beginning of the middle Pleistocene when good weather returned, the first human society was established, and the recovered abundance brought about a great explosion of joy which expressed the flowering of communal relations. And since then, at each return of spring time, people celebrate anew the birth of *Man*: the feasts and the games, the songs and the dances, bring back to life the unanimity of ancient times, the mutual reflection which envelops the whole, the joy given to man by man himself, the human social essence in its immediate simplicity, the social relation transparent to itself.

IV. THE SIGN OF THE PHALLIC WOMAN AND OEDIPAL SEMANTICS

Primitive communism, in its immediate form, made possible the first development of tool technique: during the Chellean period we notice that a perfecting of bifaces takes place. At the end of the Chellean the first improvements in flake tools appear. With the transition to the Acheulean period the oval bifaces, the first pointed implements, and scrapers appear.

Then apparently signs of an inverse evolution appear.[30]

In the first hunting camps of the early Acheulean, we notice a curious qualitative degradation of stone tools. Thus at Torralba, gross and primitive bifaces were found. And yet the inventory of animal bones shows that the hunt had made notable progress since it included such big game as the elephant or the rhinoceros. It was a big camp which lasted for quite a while, where the hunters made use of fire and carved big game. The case of the Choukoutien camp is especially typical. There, stone implements seem so amorphous that some scholars have classified it as pre-Chellean. And yet Sinanthropus knew know to use fire, and the volume as well as the structure of his skull show that human psychism had considerably developed since Pithecanthropus. The age of the bed (400 000 years) also confirms that it must be dated at the early Acheulean. Sinanthropus still remains within the limits of *Homo faber primigenius*, as one of its subspecies, but his progressive characteristics show that he is at the highest level of lower Paleolithic development.

The apparent signs of the degradation of the technique of stone-cutting – which we also find in other deposits of the early Acheulean, such as the lower bed of La Micoque, Kiik-Koba, Ehringsdorff, etc. – thus refer back not to a phase of decadence, but rather to a social crisis provoked by the increase of production itself. The undifferentiated structure of the original community began to be a fetter to the development of the productive forces. Perhaps it is to this social crisis with its ensuing conflicts that it is appropriate for us to connect also the origins of cannibalism, of which we find abundant proofs particularly at Choukoutien.

We have shown earlier that the use of fire and of little flaked tools had led to the formation of a household industry, which at the beginning of the Mousterian, in the broad sense (Riss), involved the differentiation of pairing families within the heart of the original community. It is probable that this industry began to develop during the early Acheulean in a collective form which, from that epoch, gave birth to the conflicts whose development would later motivate the transition to the endogamic community.

The household industry required particular care from the women who partook in it, which was obviously incompatible with the traditional libertarian customs of the original community. In other words, there arose a sharp contradiction between the social necessity of suspending all sexual relations with the women busy with the work of the collective household, and the unfettered liberty which until then had prevailed in those relations. The damages caused to the community by the seduction of the housewives entailed violent reactions against the guilty, which created on the whole a permanently strained atmosphere, with frequent quarrels; this is where we can find the explanation for the astonishing loss of quality in lithic tools, observed in many camps of the early Acheulean.

The conditions of the use of fire can give a particularly striking example of the conflicts which arose in that epoch. Fire required permanent upkeep, for it is probable that the men of the Acheulean could not yet produce it at will. This upkeep was obviously one of the major household tasks entrusted to women, and we can easily understand that when they let themselves be distracted by love games, the collective fire was in danger of going out, which would have plunged the whole community into a horrible catastrophe. Such distractions thus provoked vigorous reactions from the witnesses. After many experiences of this kind, a *sexual taboo* was finally decreed – perhaps the first taboo of humanity – in the form of a strict and absolute interdiction of the men accosting women and of women letting themselves be accosted by men around the collective fire-place. The rigor of such a taboo was underscored by the death penalty, and since the crime was committed around the fire, it was undoubtedly into that very fire that the guilty were thrown. And since burned human flesh probably evoked that of game, the anger of the group 'led them to eat it' – a practice which soon takes ritual form where the social interdiction imposed by the activity of production finds its *mystified sanction*. With the constitution of the *cannibal rite*, which in all likelihood appeared with the Sinanthropi,[31] and which we can consider the starting point of all *sacrificial rites of religion*, the beautiful immediate unity of the original community is itself suppressed in order to make room for this long paradoxical movement, where the development of the human social essence, founded on the progress of the productive forces and relations of production, appears in the upside-down form of its *negation* as *alienation of man from himself*; this paradox will end only in our days with the proletarian revolution and the dictatorship of the proletariat, the construction of socialism and of civilized communism, the *negation of the negation* which alone makes for the full blooming of the whole positive heritage of the historical development.

It was also undoubtedly at the early Acheulean that it is appropriate to search for the genesis of the phantasm of the phallic female, which plays a considerable role in the psychoanalytic theory of the Oedipus complex. According to Freud, the Oedipus stage in the little boy corresponds to an archaic genital organization, the 'phallic organization', which develops in children of both sexes, and is characterized by the predetermined conviction that all human beings, women as well as men, possess a masculine organ. Freud makes these two psychic formations completely coincide in time: "This phallic phase," he says, " . . . is contemporaneous with the Oedipus complex."[32]

According to various observations, however, it seems that the phallic stage would begin somewhat sooner. Thus, with respect to a little girl of two, Abraham relates that

> One day, as her parents were taking coffee at table, she went to a box of cigars that stood on a low cabinet near by, opened it, and took out a cigar and brought it to her father. Then she went back and brought one for her mother. Then she took a third cigar and held it in front of the lower part of her body. Her mother put the three cigars back in the box. The child waited a little while and then played the same game over again.

As Abraham rightly remarks,

> The fact of the repetition of this game excluded its being due to chance. Its meaning is clear: the child endowed her mother with a male organ like her father's.[33]

We can thus place the beginnings of the phantasm of the phallic female during the course of the third year of the child's development, which in phylogenesis would correspond to the Acheulean, since we saw earlier that age two corresponds to the Chellean and age three to the Mousterian.

The meaning of the female phallus is found in current language usage when, in a transparent allusion, we say of a woman that 'she wears the pants'. The phallic woman is the woman who is considered and who considers herself *as a man*. The image of the virile organ functions as a signification which tends to affirm a masculine social state. And it is this 'status of masculinity' which is signified and claimed by the appendage that the little girl mentioned earlier attributed to herself after having attributed it to her mother. As Abraham does indeed observe, "She could have noticed long ago that only her father smoked cigars, not her mother. Her tendency to place the man and the woman on an equal footing is concretely expressed by giving a cigar to her mother."

If we return to phylogenesis, we may believe that this artificial phallus was carried by the Acheulean women during their household work around the collective fire-place, in order to avoid the advances of lovers. Its meaning can be stated as follows: "This woman, in the exercise of her social functions, must be respected and considered *as a man*."

In the Acheulean bed of Burbach (Lower-Rhine), Forrer has discovered in a niche serving as a hiding-place a deposit comprising halves of premolar hippopotami teeth of a phallic appearance.[34] If our hypothesis is correct, these phallic teeth could have been carried by the housewives of that epoch as a sign of the sexual taboo which had to guarantee them peace during their work in the service of the community. Such an interpretation would find some confirmation in an observation by Abraham who relates that one of his patients dreamt that along with other women she carried a gigantic phallus taken from an animal.[35]

According to certain data based on the interpretation of dreams, the phantasm of the female phallus would appear on the navel. The Acheuleans evidently went around stark naked, as we still see today in the case of the aborigines of Australia. In that state, however, they wore a belt, and it is possible that the belt was humanity's first garment, for it can be used to hold little indispensable objects. Thus, if the Acheuleans carried their symbolic appendage at the belt, it would come precisely to the navel. We can consider it as the symbol of the good housewife, the symbol of the dignity of the mother, guardian of the hearth.

The woman's phallic symbol was probably generalized during the transition from the early Acheulean to the late Acheulean. In fact, the magnificent flowering of tool technique at the late Acheulean, which marks the transition from the lower Paleolithic to the middle Paleolithic, assumes that the social crisis had been resolved. The carrying of the female phallus, with the internalization of that symbol in consciousness, had undoubtedly reinforced respect for the taboo on housewives, which stabilized the division of labor between the sexes, thus securing progress in the development of the productive forces.

The development of production, however, brought about new relations among the producers, which generated new conflicts. It is probable that toward the end of the Acheulean, the development of the use of fire led, if not to its production, at least to its conservation under favorable conditions. Thus we still see in our day some primitive people keeping fire in wreaths of grass or bark where they can conserve it for days. Such a procedure permits individual usage, while the permanent upkeep of a fire-place is possible only for the collective group. Now the individualization of the use of fire led to the division of the collective household into pairing households whose origin was also favored by progress in the manufacture of small specialized flaked tools. As the efficiency of the pairing household was evidently superior to that of the collective household, the formation of pairing families, evidenced by the little hut foundations of the Riss, came to be imposed for reasons of economic development.

We saw earlier that the development of the pairing family under the conditions of the biological tragedy of the woman necessarily led to the prohibition of sexual relations for the entire young male generation, from puberty to about thirty years of age. It is true that it was probably also at that time that, as a compensation, orgiastic feasts were originated, whose tradition has been perpetuated among all primitive people, in order to revive a moment of the lawless freedom of the original community. But in normal times all the young men were frustrated. Now, as we can see with the typical model that we have proposed, their number was nearly equal to that of the mature men, which created an unstable equilibrium on the level of relations based on strength, and thereby a state of permanent conflict. The mature men, or '*Fathers*' undoubtedly prevailed by virtue of their physical strength and experience. But their superiority, undoubtedly sufficient to secure them priority in matters of marriage, was not such that they could impose unqualified submission upon the young men, or '*Sons*'.

Under the conditions of a demographic disequilibrium between the sexes, it was in the interest of the community that the mature men established a household rather than the young men, since they were obviously more capable of assuming the responsibilities. And the practice of marrying at about the age of thirty realized a distribution which, in fact, enabled each man to obtain a woman in his turn. From the point of view of the '*Fathers*' such a rule appeared as the sole possible manner of applying the old communal principle to the new relations created by the formation of the pairing family.

But from the point of view of the '*Sons*', the abstinence which was imposed upon them, contradicted in an intolerable way the libertarian and egalitarian traditions inherited from the original community. It is true that already at the preceding stage of evolution, a first restriction was imposed by the sexual taboo regarding the housewives. But this taboo concerned every man of the community, and the women who had completed their collective service were once again available to all. The taboo which was now imposed, on the contrary, exclusively concerned the men of the younger generation.

These young men could not understand, and even less accept, the long wait imposed upon them. In the name of the old communization of women, they became indignant about the differentiations which the new rule introduced. And in the *licentious dispute* which followed, the communal principle of old paradoxically transformed itself into its opposite, suddenly assuming the unexpected form of the *anarchical individualism of jealous desire: jealousy availed itself of an out-of-date right*, and there necessarily resulted continual conflicts, which from time to time had to end in bloody battles.

In the language of real life modeled by that social tragedy, *new linguistic gestures* were thus constituted which gave to the words 'father', 'mother', and 'Sons', already put into use in the preceding period, a contradictory semantic meaning. At the lower Paleolithic, the relation of the real mother, or of the women her age, to the son could only be a relation of rearing and education, to the exclusion of all sexual relation, since they died before the son could reach the age of puberty. At the Acheulean, moreover, the phallic symbol that the women carried during their household tasks – which included the care of children – suppressed in advance any possible sexual allusion in that relation. The gestural signifying which projected the image of the '*Mother*' for the 'Son' thus could only mean the '*helping and phallic foster-mother*' who is to be considered '*as a man*'. And it is that meaning which, still at the Mousterian, is reactivated in childhood. But reaching puberty, the young Neanderthal man, having lost all of his previous mothers – that is, his real mother and the women her age – finds himself obliged to project that status onto the women of his own generation, whom he must now consider as his '*Mothers*' since they have married his '*Fathers*' and since he, himself, will later on marry one of their daughters. And since, at the same time, he felt confident of the old law of the communization of women, and considered himself right in desiring them, there is now added, as a result, the contradictory meaning of '*Mother, object of desire*' to the primitive semantic content of the '*Mother*' as '*helping and phallic foster-mother*'. The same is true for the '*Father*' symbol, which, in the preceding period as in the childhood of the young Neanderthal man, could signify only '*educating Father*', and is now overloaded with the contradictory meaning of '*Father, rival of the Son*'. Such a formation is all the more incoherent in that it overthrew the whole of the previous semantic structure since the communal regime in the first human society excluded all forms of rivalry in general.

The endogamic community which appeared at the Mousterian thus lived in a state of *permanent crisis* perpetuated by the instability of the relation of confronting forces and sanctioned by the *semantic distortion* of language which reflected the *real distortion* of the matrimonial relations, and which in its turn was reflected in an affective deformation of consciousness.[36]

V. THE CASTRATION SYMBOL AND THE FEMALE OEDIPUS

The 'Oedipal' crisis was undoubtedly prolonged throughout the whole of the Mousterian, for it probably could not occupy such an important stage in the development of the child of our day, if it had not been fixed during a long stage of prehistory. It is true that at the late Mousterian a slight attenuation may have occurred, since we find in it traces of a certain softening of mores.[37] But an effective solution was possible only on the basis of a material change in the disposition of social forces. It would be falling into profound idealism to imagine, as in the hypothesis of *Totem and Taboo*, that the 'murder of the father' might have been followed by the 'repentance' of the Oedipal brothers. Because of the instability of the power relations between the Neanderthal '*Fathers*' and '*Sons*', victory could, depending on the situation, pass from one camp to the other. But in cases where the '*Fathers*' were getting beaten, the '*Sons*' simply took their place and just a few generations later the same problem occurred again. The crisis thus could be resolved only with the appearance of a new relation of forces, which happened only with the transition to the upper Paleolithic.

Probably because of economic development, which bettered the way of life, human longevity increased. According to Vallois's statistics,[38] in a group of 12 Neanderthal men more than twenty years old, we find that three of them are older than forty, or 25%. In a homologous group of the upper Paleolithic, among 20 men older than twenty, we find that nine of them are older than forty, or 45%. It is true that the proportions thus calculated do not refer at all to the population of prehistoric communities themselves, but solely to the fossil collections where the age of the subjects was determined only for the time of their death. The very fact, however, of the increasing number of older men is not doubtful, and we may attempt to draw more or less qualitative inferences from it.

It is probable that within the communities of the upper Paleolithic, the group of older men, while still obviously only a small minority, nevertheless became sufficiently important to constitute a particular social stratum. The adult male population is now divided into three distinct groups: the young men, from puberty to about thirty years old, the mature men, from nearly thirty years old to about forty-five; and the older men from forty-five to about sixty. This last group, even though not numerous, compensated for its numerical inferiority with its superior experience. We should perhaps attribute to this new stratum in the population the technical and artisitic progress which mark the appearance of *Homo faber sapiens*. Be that as it may, it was very probably that group which led society, an inference which is borne out by the fact that a favored gerontocracy prevails in the most primitive societies presently known. It is true that these men were not yet, in general, old men in the present meaning of the term. If their power, however, is not defined as a 'gerontocracy' strictly speaking, it nevertheless appears as the original form of gerontocracy, or even better, as a 'proto-gerontocracy' since it is based only on greater age with the difference that it was more capable of activity and initiative than would be the case in a gerontocracy in the present sense. It is highly probable in any case that these aged men, with the strength of experience and supported by the group of mature men, achieved a crushing degree of power, which enabled them to put an end to the conflict of generations, by imposing obedience on the young men.

This is undoubtedly a good place to situate the origin of circumcision, that strange operation found at the center of the puberty initiation rites among primitive people. Where it seems to be missing, it is in reality replaced by an equivalent rite, like the pulling of a tooth, the shaving of the head, etc. At the same time we will be able to shed some light on a fundamental datum of psychoanalysis, the castration complex, which until now has always been a source of embarrassment for psychoanalytic theory.

According to the authors of the *Language of Psychoanalysis,*

> The castration complex is met with constantly in analytic experience. The problem is how to account for its all but universal presence in human beings when the real threats from which it supposedly derives are far from being always evident (and even more rarely carried out!). It is quite obvious, moreover, that the girl could hardly for her part experience as serious a threat to deprive her of what she has not got. This ambiguity has naturally led psychoanalysts to look for alternatives to the threat of castration as the castration complex's concrete basis in reality.[39]

Thus Stärcke tries to find the original schema of castration in the withdrawal from the maternal breast, Rank, in the traumatic experience of birth, and Freud himself in the daily removal of the intestinal content. Actually, Freud's remark was to the point when he said that when circumcision is mentioned to children, they spontaneously interpret it as a castration. But the founder of psychoanalysis did not draw the consequences imposed by this observation. Perhaps it was retained in the theory of *Totem and Taboo* according to which the dissolution of Oedipal behavior in phylogenesis is due to a purely psychological motive: the repentance of the sons. Had he searched for the origin of the castration complex in the circumcision rite, which precisely brings the Oedipal complex to an end in the child, he undoubtedly would not have failed to see the social essence of this whole development.

The study of circumcision in classical ethnology also lost itself in the most diverse directions. Hastings's *Encyclopedia* cites more than thirteen different theories before concluding that none of them is fully satisfactory, and that in all likelihood it was probably a rite of a religious origin!

Circumcision, evidently, exists among many people only by virtue of survival, and under these conditions it is subjectively justified for the most varied motives. But if we look at the Australians, where it still seems to be close to its origin, we may observe that those who practice it have a perfectly clear awareness of its real significance. Thus according to Strahlow, the Arunta explain the reason for this custom in the following manner:

1. This painful operation must subjugate to the authority of the older men the young men who until then were not accustomed to such obedience. They must now obey *knaribatana kankueritjika* the old men. The reason for this is indicated in the legend of the two Sparrow-hawk men, Lakabara and Linjalenga, who insisted that the men strictly observe that custom, and warned them that the boy who did not go through that rite, would become an *erintja*, who will kill the men of his own tribe and eat them.
2. Circumcision must set up a barrier against the exuberances of the young pubescents.

Strahlow also emphasizes the fact that

the interest of the older men gave them a weighty motive to carefully keep up that practice. On the one hand, it imposed on the young men an obligatory tax, *tjaneritja* to be paid to the elders, generally in the form of game. On the other hand, the rules concerning food forbade the youth to taste the choice pieces under any pretext, since these were reserved for the elders. Finally 'in the good old days', the young men had to wait for their first grey whiskers to grow before they could possess the women that they were promised, or else they were given older women, while the old men took for themselves the privilege of marrying as many young women as they wanted.[40]

We know that as the organism ages it is in need of both richer and more delicate food. And we may believe that the men of the Paleolithic, due to the hard life they were leading, aged faster than the men of our day. For the new stratum of older men which appeared in the transition from the Mousterian to the upper Paleolithic, it was probably already a vital problem to get hold of the better pieces of game. They could only do so by securing the total submission of the young, which assumed the dissolution of the 'Oedipal' crisis.

We have just seen that the meaning of circumcision as practiced by the Australians consists in making the young pubescents understand that they must obey the old ones in all things, and in particular must renounce sexual satisfaction until they reach maturity. The rite is thus a *sign*. And if we examine the significance of that sign, we can easily see that it has the form of a castration: in a very exact way, it is a *mitigated castration*, as we can verify by reading the detailed description given by Spencer and Gillen.[41] Moreover in the legends of Alcheringa concerning that ceremony, we find precisely scenes where an actual castration is substituted for the expected circumcision.[42]

Circumcision is thus an attenuated castration or a symbolic castration, a symbol which projects as meaning the threatening image of a real castration. The symbolic form, of course, derives from the actual form. It is probable that from the beginning, during the Mousterian, such castration sanctioned the interdiction of semantic incest with the '*Mother*'. At that epoch, however, since the '*Fathers*' were unable to impose their authority fully, the prospect of punishment did not scare the disobedient '*Sons*'. And it is precisely an attitude of this kind that we meet again in the child of three to five years old, at the *age of Neanderthal man*: "The boy," says Freud, "does not believe in the threat or obey it in the least."[43] It is only with the upper Paleolithic that, thanks to the action of the new stratum of older men, the power of the '*Fathers*' will crush the resistance of the '*Sons*'; the castration symbol, systematically applied in the form of circumcision to all young pubescents, made them renounce all sexual desire until the age required for marriage. And this is precisely what is reactivated in the child around age five with the castration complex: The boy, says Freud, "[is forced] to believe in the reality of the threat which he has hitherto laughed at,"[44] and this threat literally explodes not only the Oedipus complex, but the whole infantile sexual organization as well since the child enters into the 'latency period' which must last until puberty.

"The Oedipus complex," says Freud, "is . . . a phenomenon which is determined and laid down by heredity and which is bound to pass away according to programme when the next pre-ordained phase of development sets in."[45] The notion of 'predetermined program' can be understood in two senses: either the program would be nothing but the spreading out in a temporal form of a, so to speak, *a priori* structure, namely for the Oedipus complex, the altogether general opposition between Desire and Law, an opposition resolved by the castration complex which incarnates the prohibitive function of the law and guarantees its execution. Or, on the contrary, the program reproduces in ontogenesis the historical dialectic of phylogenesis, so that each of its points can have only a specific and relative value, historically conditioned and not general and absolute. Now if the castration complex has for its function the actualization of a structure which is itself atemporal, it should put an end, in today's child, only to his desires for his mother or her sister, since these desires are the only ones forbidden in our society. In other words, nothing in principle would prevent the little boy from transferring his love life outside of his family, since at the age of five his social relations already normally transcend the strictly familial framework. Now, experience shows that this is not the case. The child does not transfer his sexual impulses to a girl friend, he purely and simply suspends them. The fact of the latency period shows that the castration complex does not play for him the same structural role of sexual relations on the human plane. This structural role will be realized at puberty by the action of existing social rules, and if the phantasm of castration then returns, it can only be as a survival from infantile prehistory, and play an altogether subordinate role.

The castration complex which exerts its specific action at about age five is but the reactivation of the symbolic castration imposed by the '*Fathers*' of the upper Paleolithic, so as to suspend completely the whole sexual life of the male youth until maturity. And since the castration symbol had for its essential function the guaranteeing of the '*Fathers*'' power over the whole community, it also logically had to apply to the young nubile women. We undoubtedly could take up again here the objection of the authors of *Vocabulaire de la psychanalyse*: "It is rather obvious that the girl could not possibly see herself really threatened by the deprivation of something she does not have." However, we have seen earlier that the phantasm of the female phallus, eliminated by the little girl's castration complex, is itself but the memory of a very real object; the symbol of the phallic woman carried by the housewives of the Acheulean. It was probably still in use during the Mousterian in order to avoid the advances of the young coxcombs. The regular carrying of such a sign was evidently reflected in consciousness by the representation of the woman as identical to man, and such a representation produced in the women of the epoch a 'masculinity complex', which is reactivated in the little girls of our day during the phallic stage.

Now, when the older men of the upper Paleolithic assumed the leadership of the community, the masculinity complex of the women could not be allowed to become an obstacle to the establishment of the proto-gerontocracy. In order to obtain their submission, the old men found that the easiest way to do this was to suppress the very sign which produced the fundamental meaning of that complex, and not only the material sign but also its ancestral memory: the phantasm of the phallic woman. The castration symbol was thus generalized to include the young nubile women, and this is what we rediscover in the defloration rites still in use by many peoples. The ritual defloration comprises not only the introcision of the hymen, but also often the excision of the nymphae [labia minora] and sometimes of the clitoris, and this excision plainly reveals the meaning of the original rite, as symbolic castration. It was a question of systematically taking away from the young girls the phallic phantasm inherited from Acheulean and Mousterian times, and thus eliminating along with the masculinity complex, any temptation to resist the new power.

We see that the castration symbol presents a different meaning here from the one it has for boys. While for the young pubescent boy it represents a symbolic ablation, or threat of ablation, of his real organ, for the young nubile girl, it somehow realizes an actual ablation of an imaginary organ. And this is very exactly what is reactivated in the little girl of our day at about five: she considers her real anatomy the result of a castration. "The girl," says Freud, "accepts castration as an accomplished fact, whereas the boy fears the possibility of its occurrence."[46]

In the typical model that we have proposed for the distortion of matrimonial relations during the Mousterian, we have assumed, for the sake of exposition, that all the young girls reaching nubile age married the young men of 29. Actually, the process is more complex. If, in fact, we admit that each year there is an equal number of boys and girls reaching puberty, since the demographic disequilibrium between the sexes, due to the biological tragedy of the woman, begins only with the first childbirth, as a result the class of young men of 29 must clearly be less numerous than that of the girls of 14, since from puberty to age 29 there had to occur a certain number of deaths among the young men. There is thus each year an excess of young nubile girls, who were married by the widowers of the women who generally died before thirty years of age. For the young married couples the difference is rather minimal, since the widowers were in the prime of life, being generally less than forty-five years old.

But in the upper Paleolithic, the situation appears to be entirely different. In effect, longevity having increased, the widowers had more chance of finding themselves among the older men. And in a gerontocracy, these old widowers clearly had priority in being awarded the excess young nubile girls. As a result these young women married men three times older than they, who in fact belong to the generation of their *fathers*. It is true that during the Mousterian, the mature men between thirty and forty-five were already called '*Fathers*'. But in reality they did not appear to be much different from the young men of 29, and their second marriages, in case they had become widowers, with young nubile girls, did not create any particular problem. Now, on the contrary, the widowers who remarry are already men of an advanced age, and it is even the oldest widowers who exercise priority in the selection of young nubile girls by leaving any [older] widowed women to the others. We may believe that at the beginning, the young brides manifested their discontent with their fifty-year old husbands. And this is undoubtedly one of the reasons which led the proto-gerontocracy to dissolve the phantasm of the phallic woman by the systematic application of the castration symbol to all young nubile girls. For evidently the masculinity complex which rested on that phantasm crystallized the discontent of the young women obliged to live with old spouses.

But once the rite is established the situation becomes reversed. For the mutilation, entirely real and painful, which took away the imaginary organ from the young nubile girls, also took away their spirit of revenge. And as the older men, having firmly established their power over all the youth by means of the castration symbol, now enjoyed an uncontested prestige, the young women married to widowers whose respect increased with their declining years, could not help but be seduced by such a flattering social position. Thus they came to love their old husbands, who were of the same age as their fathers. In other words, they loved their '*Fathers*'.

"*Whereas in boys the Oedipus complex is destroyed by the castration complex*, in girls it is made possible and led up to by the castration complex." Freud considers this difference as "an intelligble consequence of the anatomical distinction between their genitals and of the psychical situation involved in it."[47] Clearly, anatomy, without being 'destiny', is always the basis, but it could not determine a 'psychical situation' without the mediation of social relations. And the difference considered here is essentially of social origin and not an anatomical one.

In the process we have just described in the upper Paleolithic, the old men who married young nubile girls were widowers. The young spouse in her household with her old husband thus did not confront any rival. We know, however, that from the Mousterian to the upper Paleolithic, the biological tragedy of the woman was notably attenuated, since the sex ratio fell from 163 to 141. Now, the matrimonial system which we have described, was adapted to the demographic situation of the Mousterian, and consequently, there is now a relative excess of female population. In other words, the longevity of the women increased more than that of the men, so that there were fewer widowers, and consequently there still remained a certain number of young nubile girls once all the widowers were remarried. According to the rule we have indicated, the oldest marry the young nubile girls and the younger men the widows. Under the conditions of the domination of old men, there was evidently no question of lowering the marrying age for the young men. The '*Fathers*' rather exploited the situation in order to establish polygamy in their favor, as we can still see in the Australian tribes of the nineteenth century. In short, the *Elders* have taken the excess young nubile girls as second wives, and consequently in polygamous families, where the husband and his first wife are of the age of the father and mother of the new spouse, she can easily become jealous of her '*Mother*' with respect to their common spouse, who for her occupies the position of '*Father*'.

The Oedipus complex is thus constituted. However, because of the very conditions of this constitution, it could not have the sharpness of its male predecessor. In effect, we have just seen that the love of the young woman for her old husband did not originate from sexual desire, but simply from the prestige which crowned the *Elders* with a halo as holders of power. The rivalry between the '*Daughter*' and the '*Mother*' for the favors of the '*Father*' having only a social foundation, thus could not generally end in tragic consequences – all the more since there was no question of the '*Mother*' forbidding any sexual satisfactions to her '*Daughter*' as the Neanderthal '*Father*' used to do with respect to his '*Son*'. The Oedipus complex is thus here little developed, and this is exactly what we rediscover in its reactivation in ontogenesis: "The girl's Oedipus complex . . . ," says Freud, "seldom goes beyond the taking of her mother's place and the adopting of a feminine attitude towards her father."[48]

During the upper Paleolithic, the *bio-social evolution* comes to an end to make way progressively for *social history*. Consequently the female Oedipus complex, constituted under the conditions of the upper Paleolithic, can be considered as the last *archeo-psychic* formation which is reactivated in a predetermined manner, because of the laws of maturation, in infantile prehistory. The following forms of language and consciousness which we may characterize as *paleopsychic*, no longer belong to bio-social evolution, but to the first steps of social history, and are constantly preserved, not in the biological memory based on the enduring of the dialectical moments of phylogenesis in the maturation of the child, but in the social memory based on historical documents. They must therefore be reactivated essentially by education, which mediates the constitution of the *neo-psychic*.

We thus understand that the dissolution of the female Oedipus complex does not involve any predetermined crisis as is the case for its male counterpart. "In girls," says Freud, "the motive for the demolition of the Oedipus complex is lacking." "One has an impression that the Oedipus complex is . . . gradually given up . . . "[49] Actually, the female Oedipus complex is progressively replaced in the psychism of the child by the educational forms of social relations, for the very simple reason that the infantile sexual organization has had its day: *The program has ended*.

We have admitted that the double circumcision rite, male and female, which is reactivated in the castration complex, originates at the beginning of the final phase of bio-social evolution, or at the beginning of the upper Paleolithic. It would obviously be a good idea to verify this hypothesis by archeological and ethnographic documents.

We can find an indirect testimony of the female castration symbol in the style of the Aurignacian Venuses,[50] for such an exaggeration of the specific shapes of the woman undoubtedly would not have been possible if the phantasm of the female phallus had not been eliminated in the society of that time. More precisely, by forcefully insisting on the female attributes, the artist thereby exorcised once more the obsessive symbol of the phallic woman and confirmed its ablation, which in the social consciousness reinforced the new power of the gerontocracy. And since it is likely that female circumcision was conceived by analogy with its male counterpart, it must consequently date from about the same epoch.

The Australian myths make circumcision go back to the very origins of humanity. Thus, according to the Arunta tradition, that rite was instituted by two 'self-existing' beings, the *Ungambikula*, who, living in the western sky, could see on earth a number of *Inapertwa* creatures, that is, rudimentary human beings whom it was their mission to transform into men and women.

In those days there were no men and women on earth, but only the *Inapertwa* who were varied in appearance and dwelt in groups along the shores of the salt water. They had no distinct limbs or organs, but appeared in vaguely human forms all doubled up into a rounded mass in which just the outline of the different parts of the body could be seen.

Coming down from their home in the western sky, the *Ungambikula* took hold of the *Inapertwa*, one after the other. With their stone knives or Lalira they separated the arms; then the fingers appeared when four clefts were cut at the end of each arm; then legs and toes were added in the same way. The figure could now stand, and after this the nose was added and the nostrils bored with the fingers. A cut with the knife split open the mouth, which was opened and shut several times to make it flexible. A slit on each side separated the upper and lower eyelids, hidden behind which the eyes were already present. Another stroke or two completed the body, and thus, out of the *Inapertwa*, men and women were formed.

These *Inapertwa* creatures were in reality an intermediary stage in the transformation of various plants and animals into human beings, so that the newly made men and women remained intimately associated with the ancestral species from which they originated. Consequently each individual necessarily belonged to a totem which represented precisely its animal or plant origin. Tradition cites six totemic groups constituted in this way: The Plum Trees, the Grass Seeds, the Large Lizards, the Small Lizards, the Parakeets and the Small Rats.

All except the Plum Tree men were circumcised by the *Ungambikula* by means of a fire-stick, *ura-ilyabara*.

The *Ungambikula* refused to circumcise the Plum Tree men because of some *Oruncha* devils who killed and ate almost all Lizard men and women, and also a great number of people with other totems. Two Lizard men who survived the slaughter armed themselves with heavy spears and set up an ambush in a narrow gorge. When the *Oruncha* appeared they killed them.

Spencer and Gillen's account, which we have just summarized, concerns the first period of the *Alcheringa*, the time of the ancestors of the Arunta tribe. The *Alcheringa* comprises four periods. In the second, the work of the *Ungambikula* was followed by the men of the totem of the Little Falcon, who introduced an important innovation by performing circumcision no longer with a fire-stick *ura-ilyabara*, which caused many fatal accidents, but with a stone knife *Lalira*. In the third period of the *Alcheringa* the rite of sub-incision appeared. Finally, it was only in the fourth period that the rule of exogamy was established.[51]

We have spent some time on this remarkable tradition of the Aruntas not only because it represents a first effort to conceive in a more or less coherent form the process of anthropogenesis, but also because it testifies to relatively precise mementos of the origins of society.

According to Kabo's investigations, the population of Australia began at the beginning of the upper Paleolithic with the immigrants who came from New Guinea or from the Sunda Isles. At that time, Australia was connected with the Asiatic South-East, by a string of isles sufficiently close to one another to enable a continuous passage in these regions with the primitive means of navigation at the disposal of the first representatives of *Homo faber sapiens*. A prehistoric site such as the Koonalda Cave, in the Australian interior, already dates back 31 000 years. The arrival of immigrants on the northern coast of Australia thus could only occur at the beginning of the upper Paleolithic. It cannot be any earlier since we have been unable to find any traces either of the Archanthropi or of the Paleonthropi on the Australian continent.[52]

The *Inapertwa* who, according to the tradition, dwelt in groups along the shores of salt water in the first period of the Alcheringa thus represent the first proto-Australian communities settled on the northern shore at the beginning of the upper Paleolithic. They originated from the endogamic communities of the Mousterian, and now underwent a fundamental reform, which transformed them into a totemic community, always endogamic of course, since exogamy was established only in the fourth period of the

Alcheringa. We should notice, moreover, that the only woman mentioned for that first period is a Lizard woman, the wife of a man of the same totem. We have here a precise indication of the existence of the first endogamic stage of totemism.

The reform appears as the creation of a new rite which we may call the '*rite of hominization*'. The novice *Inapertwa*, curled up like an embryo and somehow considered as a chrysalis coming from an animal or a plant, was subject to a series of operations which were intended to give him human form, following which he would acquire the quality of man in the sense of *social man*, in other words the *status* of membership in a totemic group. It is probable that already during the Mousterian the endogamic communities identified themselves with animal or plant species, which was a way of affirming their particular unity by differentiating themselves from one another. We can give to this first organization the name of pretotemic, which, on the internal plane, remained as we have seen, more or less anarchistic because of the unstable equilibrium of the social forces within the community. The *Inapertwa*, intermediary beings originated from plants or animals, represent precisely the *dialectical moment* of the pretotemic community, which, having arrived at the end of its development, must pass to the real totemic organization. The latter implies a *hierarchized unity* under the authority of the Elders, the '*Fathers*' of more than forty-five years of age, whose power will be idealized precisely in the representation of the *totemic Ancestor*. In pretotemism, which bore the mark of Mousterian anarchy, the identification of the community with an animal or plant species still implied only an undifferentiated identity, so that this species was not yet represented as the Ancestor. The mystical relation to the totemic Ancestor can be constituted only on the real basis of the *Elders*' power whose totemism appears precisely as the *religious ideology*. At the beginning of the upper Paleolithic, however, at the original moment when the proto-gerontocracy asserts itself, the totemic ideology is not yet fully constituted. The authority of the *Elders* here refers not to the totemic Ancestor, but to the transcendent might of the two 'self-existing' *Ungambikula* descended from heaven. And it is only at the end of the '*rite of hominization*' that the ideal relation to the mystical Ancestor, animal or plant begins to be asserted. This relation will only assume its full ideological value only later, when the gerontocratic regime is firmly established. At the point at which we have arrived, the totemic ideology has only just been born, and its birth is precisely this '*rite of hominization*' of which the Arunta have preserved the memory.

At the beginning of the ceremony, the *Inapertwa* novice appears as the symbol of a pretotemic community whose already mature state of evolution is indicated by the chrysalis form reached by the animal or the plant with which that community identified itself, a form where the human traits are vaguely outlined. Then the *Elders* appear who, officiating in the name of the 'Self-existing' *Ungambikula* descended from heaven, transform the chrysalis into a man with a stone knife. The rite thus appears as a *complex of signs* which make the novice understand that it is precisely to these *Elders* that he owes his human existence as such, and that he consequently has the duty to submit to their will. It is probably on that occasion, as is the custom in initiation ceremonies in general, that he is taught the rules of his new status. In short, the '*rite of hominization*' is the founding rite of original totemism, as the ideological justification of proto-gerontocracy.

However, if the power of the *Elders*, sanctioned by that first ceremony, could, in all probability, be imposed without any particular difficulties in the economic life, it certainly met with strong resistance in the matter of sexual problems. For in this domain, as we saw earlier, the older men awarded themselves considerable privileges, notably by the establishment of polygamy, thanks to which they hoarded the excess of young nubile girls, instead of lowering the marrying age of the young men as would have been logical. And as the youth could not help protesting against such encroachments, the *Elders* reacted by establishing the rite of circumcision as initiation, as a *preventative* symbol of castration.

We showed earlier that castration had to be practiced during the Mousterian, as a punishment for the infractions of the taboo of semantic incest with the '*Mothers*'. It is probable that in the majority of cases this punishment was inflicted only in the attenuated form of circumcision. We can find a confirmation of this hypothesis in the Arunta tradition concerning the origins of this last operation. In fact, the myth specifies that the *Ungambikula* circumcised the members of the totemic groups that they had just constituted with a fire-stick *ura-ilyabara*. Now, from the beginning of the upper Paleolithic, *Homo faber sapiens* possessed excellent stone blades and the myth moreover mentions that the *Ungambikula* accomplished what we have called the '*rite of hominization*' with their stone knives *Lalira*. Furthermore we know that the use of the fire-stick used for circumcision often caused fatal accidents, and it is precisely for that reason that in the second period of the Alcheringa it was replaced with a stone knife. The fact that the *Ungambikula*, in other words the officials of the upper Paleolithic, used the first instrument for circumcision, in spite of all its drawbacks, and not the second, which they already had at their disposal, shows that it is a question here of a *ritual survival*. As Spencer and Gillen remark, the same phenomenon will recur, when the stone knife continues to be used in various rites, when the use of iron was already known in economic life.

In our case, it is probable that the fire-stick was used during the Mousterian, because at that epoch sufficiently sharp stone knives were not yet available. Perhaps it even could be traced back to the Acheulean where we saw that the punishment which sanctioned the taboo of the housewives, was probably inflicted by fire.

Circumcision thus dates at least from the Mousterian. But at that epoch it was still only a *penal rite* which punished real infractions of the rule. It was only at the beginning of the upper Paleolithic that it became systematically generalized, as a warning of a threat of castration, and thus became an *initiation rite* . If we believe the myth, such an innovation did not go on without meeting with violent resistance. The *Ungambikula*, in other words, the party of *Elders*, did not succeed in imposing it in the totemic group of the Plum Tree, because of certain *Oruncha* devils. These probably represent the young rebellious men who took the offensive and slaughtered the partisans of that new rite, not only among themselves but even in other totemic groups. This suggests to us that these groups were already more or less connected with one another. Without clearly raising themselves to the tribal form, since exogamy did not yet exist, they undoubtedly already had sufficiently close relations in order to enable the *Elders* to bring assistance from one group to another. And this is what explains the generalized offensive of the young Plum Tree men.

Circumcision, as we have seen it constituted as an initiation rite of endogamic totemism at the beginning of the upper Paleolithic, does not evidently function as an interdiction symbol of incest in general, but only of the incest of the '*Son*' with the '*Mother*'. On the contrary, it opens the way for the incest of the '*Father*' with the '*Daughter*'. And it is precisely in that form that it is reactivated in the last period of child prehistory, since the castration complex, on the one hand 'explodes' the male Oedipus complex, but, on the other hand, makes possible and introduces the female Oedipus complex. Incest in general will be forbidden only with the establishment of exogamy. Circumcision, which is preserved in exogamic societies, then takes on the meaning of a total interdiction of incest. In that form, however, *it is no longer part of the bio-social evolution*. In fact, exogamy was in all likelihood established only during the Mesolithic when the bio-social evolution had fully ended.

It is obvious, in fact, that the matrimonial exchanges which probably began more or less sporadically in the upper Paleolithic, with the relative excess of young nubile girls which appeared in the endogamic communities, could develop into a regular and total form only when the equilibrium between the sexes was entirely reestablished in the population. As long as the biological tragedy of the woman continued, and as it evidently acted in a more or less uneven way according to the times and the communities, these communities never had, in general, the same sex ratio. It thus normally happened that of two neighboring communities one of them had relatively more women than the other. And since the general scarcity of women determined their price, it is rather clear that the community which had the most women, could not consent to an even exchange. In other words, even in the final phase of the biological tragedy of the woman, which extends to the first part of the Mesolithic, the matrimonial exchanges could still develop only in a *partial* form. It is thus only during the Mesolithic, when the demographic equilibrium is entirely reestablished between the sexes, that the matrimonial exchanges could be *entirely* organized, as was to be the case during the exogamic regime.

We should also notice that this complete system of matrimonial exchange presupposed a certain level in the development of the social relations between the communities based on regular economic exchanges, which is precisely what happened only during the course of the Mesolithic.

Finally, if we return to the Arunta tradition, we see that it places the establishment of exogamy in the fourth and last period of the Alcheringa. Now, according to Kabo's estimates, the Australian tribes of the nineteenth century were at the development level of the final Mesolithic. Since the first period of the Alcheringa dates from the upper Paleolithic, it seems reasonable to place the fourth and last period during the course of the Mesolithic.

Now, during the Mesolithic, with the end of the biological tragedy of the woman, the bio-social evolution was completely terminated, and we can consider precisely the establishment of exogamy as *the first step of social history*. In other words, this even could not have left any particular mark in the hereditary stock so that it is not part of the maturation program of the child. The castration complex which appears in a predetermined manner between the fifth and sixth year of the child of both sexes, can thus refer only to the double circumcision rite of the upper Paleolithic, as interdiction symbol of maternal incest and castration of the phantasm of the phallic woman, and not at all to that of the second part of the Mesolithic, which forbids all incest in general, with the father, the mother, the brother and the sister. In short, the castration complex is solely a formation of child prehistory as reactivation of the bio-social dialectic of anthropogenesis, which stops before the establishment of exogamy. And since the exogamic structure constitutes precisely the strictly human structure – in the sense of historical humanity – of sexual relations, consequently, *the castration complex as such absolutely can not play a role in the structuring of sexual relations as human relations*. The fact is obvious in itself for the little girl, since she engages herself in paternal incest precisely because of the castration complex. And it is also certain for the little boy, for if it is true that the castration complex brings an end to his maternal incest, it does not in any way bring any solution to his love life, since it simply introduces the period of latency. Now the human structuring of sexual relations, as it takes place in exogamy, is not limited to that negative form which is the interdiction of consanguinity relations: it implies essentially a *positive* orientation, by the opening up of the field of relations outside of the familial community.

VI. FROM THE NEANDERTHAL 'OEDIPUS' TO THE INFANTILE OEDIPUS

We have shown that all psychic formations connected with the Oedipus complex, which follow one another according to a predetermined order from the third to the sixth year of the child's development and which as a whole we may call the 'Oedipal formation', that is, the phantasm of the phallic woman, the male Oedipus complex, the castration complex and the female Oedipus complex, are none other than semantic forms originally produced in humanity's prehistory by the material motion of gestural-verbal signification modeled by the bio-social dialectic of anthropogenesis. And since this whole epoch is characterized by an intense evolution of the brain, the ensemble of these signifying forms, integrated in the cortical structure, have been placed on in the hereditary stock. When, as the Oedipal formation progressively developed, its obsolete elements left the social scene, and when it itself disappeared entirely from historical reality with the foundation of exogamy – since the signifying elements which survived, like circumcision, were integrated in entirely different structures – the 'Oedipal' forms preserved by heredity passed to the recessive state, *sedimented* in childhood life and only recur occasionally in the adult in the dreams of normal people and in the obsessions and hallucinations of neurotics.

Now, in its revival in the child, the Oedipal formation is confronted with a new situation. In effect, the child in his family deals with his real parents and not simply with his social parents like the Neanderthalian 'Oedipus'. It is thus to his real mother and father that he applies the distorted meanings originally applied to the social *'Fathers'* and *'Mothers'* of the Mousterian, which clearly multiplies *their traumatic resonances*. The child thus becomes paradoxically an Oedipus in the classic sense. And it is in that monstrous form that the affective and semantic distortion inherited from *Homo Neanderthalis* comes back to haunt the dreams of *Homo sapiens*; and intruding itself in the various shifts and evasions of group conflict in tribal societies and in the class struggle of the first civilizations, it conceals these real contradictions with obsessive mythical figures, like the famous and typical hero of Greek tragedy, which brings its most heart-rending characteristics to their culmination.

The Oedipal crisis which appears as a fully determined phase in human development actually does not contain a profound logic at all. For the fundamental development of anthropogenesis, as the process of the animal becoming human, essentially comprises only the acquisition of bipedal gait as a condition of the regular use of the instrument, the development of labor, of language and consciousness with the correlative advance in brain development and of social relations which in their turn condition the new developments of production. The biological tragedy of the woman, which at the time of the appearance of the pairing family, provoked the Oedipal distortion of the matrimonial relations, of language and consciousness, was in no way an essential condition, but was rather an anomaly of the development, due to the particular position of the *ischium* between the locomotor apparatus and the genital apparatus, so that the social development of bipedal gait entailed as an unexpected consequence the narrowing of the pelvis, which limited the possibilities of dilatation of the birth canal. We have seen that this anomaly, through the fetation which resulted from it, played the role of a catalyst which considerably accelerated the course of anthropogenesis. But in itself, it consists only in a maladjustment which should be and was in fact corrected, as the genital apparatus finally harmonized its function with its new morphology.

The Oedipal crisis which originated from this temporary maladjustment, and which had to find its final solution in the reestablishment of the equilibrium between the sexes in the population, can thus absolutely not be considered as a 'necessary stage', in the sense that it would impose itself as an essential moment in the development of the genus *Homo*. For it is rather clear that such a distortion of social relations and the life of consciousness cannot bring any contribution, unless a purely negative one, to the formation of human society and the human psychism.

The Neanderthal 'Oedipus' was thus in fact only a socio-psychic anomaly, which necessarily flowed from the conditions of the biological tragedy of the woman, and was thus an *historically determined anomaly*. But such a determination, in itself purely historical, had only the sense of a *pathological determination*, due to the intervention of a biological anomaly, and if it figures as a *dialectical moment* of anthropogenesis, it can only be as *a determined abnormal moment in historical dialectic*, and not in the least as a moment or necessary determination in the dialectical logic of the human development.

If we now go on to the child Oedipus, we see that he brings that anomaly to a teratological [monstrous] form, by applying to his real parents the affective and semantic distortion which in fact originally applied only to his social parents. Thereby ends the *tragic fatality* which from its origin marked the Oedipal destiny.

Of course, such a total deformation of elementary meanings, due entirely to extrinsic conditions, cannot play any positive role and consequently does not have any rightful place in the human development of man's young. We can consider it only as a *hereditary aberration* stemming from the Mousterian crisis, which is more or less reproduced in every child at a certain age because of genetic laws; in no way can it be considered as a 'general structure' defined independently of history by the triangle of 'sexual relations' of the child to his begetters. It is true that sexual sensations exist in the child. But in themselves they are blind, all the more since the organs are not mature, so that at that age, the relation to the object is possible only with the application of a predetermined meaning, which consequently implies the linguistic structure inherited from the social development in phylogenesis. The 'sexual relation', in the sense of object-relation, thus presupposes the semantic relation, and it will be enough to examine the Oedipal relation on the semantic plane to see that it does not in any way contribute to human development, except to hinder it.

Psychoanalysis has accumulated important and positive observations on the Oedipus complex and the disastrous consequences of its insufficient dissolution. The psychoanalytic cure, whose efficacy in certain cases has been demonstrated, implies some [kind of] scientific basis, and Lévi-Strauss was undoubtedly too severe when he suggested its assimilation to a resurrection of Shamanism in modern civilization. The fact that psychoanalysis cures solely by means of speech does not necessarily mean that it uses magic. Since the illness in question consists in the exaggerated development of a hereditary semantic distortion, it can be defined as an *illness of language* which implies the possibility of a treatment by language itself. We have seen that the castration complex which normally eliminates the Oedipus complex, itself consists in a *linguistic sign*, namely the threat of castration, a sign which was generalized by the proto-gerontocracy in prehistory in the form of circumcision. Since this brutal dissolution leaves pathogenic traces, it is logical to believe that it is susceptible of a correction by the use of a more supple language and an appeal to reason.

We see that the historical explanation of the Oedipus complex as the survival of the semantic distortion of Mousterian times fully accounts for the *practical results* as well as for the *factual data* of psychoanalytic observation. On the other hand, it is entirely incompatible with the *Freudian theory* and more particularly with the theory of instincts [*Trieb*]. The notion of instinct defined as "an exigency of work imposed upon the psychic in consequence of its connection with the corporeal," posits between the biological and the psychic a relation which claims to explain the second by means of the first ('in consequence of'), but actually remains purely formal and indeterminate: 'an exigency of work imposed'. In fact, since the 'work' spoken of here, does not appear to have any specific determination, in order to explain the relation in question, there remains in fact only the pure vacuum of the voluntaristic form: 'exigency imposed'. In other words, under the cover of an apparently scientific referrence to biological objectivity, this fundamental notion of Freudian theory actually refers only to the lived subjectivity, taken in its most indeterminate form, since one absolutely does not know what the proper nature of that 'exigency' is, nor how it could 'impose' itself. The reference to the biological here is only a pure postulate posited for the sake of Freudian discourse: "The forces which we assume to exist behind the tensions caused by the needs of the id are called *instincts*."[53] And it is this simple postulate which enables us to give a verbal explanation to any psychic phenomenon whatever just by adding an adjective which merely repeats the distinctive character of the phenomenon to be explained. For example, it will be said that the act of love comes from the '*sexual* instinct', hatred from the '*sadistic*' instinct, etc., exactly as in other times the sleeping effect of opium was explained by its dormitive virtue. Such a metaphysical procedure, erected into a total system of explanation, could not fail to end up in myths, and this is what Freud, with that singular lucidity that characterized him, had finally to admit: "The theory of the instincts is so to say our mythology. Instincts are mythical entities, magnificent in their indefiniteness."[54]

In opposition to the Freudian theory which contents itself with purely descriptive general notions, elaborated, to be sure, in a rich fabric of concrete facts, but incapable of grasping them in their *concrete determinations*, historical materialism has the necessary scientific concepts at its disposal to enter into the determinate content of psychic phenomena and to give them a precise explanation. From this point of view the Oedipal conflict is no longer reduced to the abstract and metaphysical opposition between 'desire' and 'law' or even between the 'individual' and 'society'. It refers to the dialectical contradiction, historically determined, between *two laws*: on the one hand, the *primitive law* of the communization of women which, in the amorphous and undifferentiated state of the first human society, during the Chellean, guaranteed, by the strict interdiction of all jealousy, the necessary unity and solidarity for the beginnings of tool production. On the other hand, there is the *new law* of the pairing marriage, imposed by the progress of productive forces and more particularly by the development of the household industry of the Mousterian. The old communal right to sexual freedom without restrictions became a hindrance to the development of the productive forces and lost all social justification, merely appearing as a simple *individualistic claim*. Now, such a claim which, in other circumstances, could have been limited to particular cases – which would not have left any trace in heredity – was aggravated under the conditions of the *biological tragedy* of the woman, and was embodied in the language of real life in the generalized form of undisciplined competition among frustrated youth. The contradiction of the two laws assumed the anguished form of sharp conflict between generations, and developed into a gigantic *social tragedy* which is perpetuated today in the *psychological tragedy* of the child and the neurotic. Thus in its human sense, the psychic cannot be directly understood in terms of the biological. It implies the mediation of social determining factors, based on the historical dialectic of production forces and productive relations, and developed in language.

By explaining the Oedipus complex in terms of the purely individual game of the child's instincts toward his 'primitive object', namely his parents, Freudianism erects into a 'general structure' of human development what in actuality is only an historically conditioned excrescene transmitted by heredity. He thus attributes a central role to this anomaly, and paradoxically gives it some sort of justification, at least on the individual plane, since its dissolution comes about only through the brutal action of a purely repressive law. There results from it an unexpected promotion, where the monstrous jealousy of the little Oedipus is raised to the dignity of a universal prototype of desiring humanity, where the bourgeois conception of human relations is made eternal. One is tempted to say that the Oedipus complex, with all of its negative aspects, nevertheless defines a transitory stage of development *as a stage to be transcended* and, eventually, in fact transcended. In other words, the human personality was constituted, so to speak, as a '*forbidden Oedipus*'. Actually, we know very well that the first formation of the personality which is characterized by the use of the first person pronoun, occurs towards age three, exactly *at the very moment of the appearance of the Oedipus complex*. There results from it that well-known ambivalence in the relation to the father. On the one hand, as a little Oedipus, the child perceives his father as a rival to be eliminated, but on the other hand, in the activity by which he is raised for the first time to the personal consciousness of self, he takes his father as a model with whom he identifies.

We obviously have here the continuation and the development of the Oedipal relation, or "an identification of an affectionate sort with the boy's father, an identification which is still free from any sense of rivalry in regard to his mother."[55] The formation of the personality thus in no way first goes through the Oedipal complex and then through the stage which interdicts it. Actually, we are dealing with two parallel paths, which are strictly antinomical, the first preceding the second one: on the one hand, the path of *affectionate identification without rivalry* with respect to the parents, which begins with the appearance of the object-relation at about age two and develops into the first form of personality at about age three; on the other hand, the Oedipal path of *jealous desire* which appears only at about age three. Starting at age three, the *child simultaneously covers both paths*. But it is obvious that the first is the only one which has a real future, and thus effectively brings about the development of a mature human personality. The second is from the beginning only a deviation and an impasse, which ends definitively with the castration complex, which reduces the Oedipus complex to the underground level of the *unconscious. The unconscious is generally the sedimented residue of the language of the transcended stages of human development*. From this point of view, we can say that the unconscious *supports* the conscious and serves as preparatory stage for its activity. But the *unconscious in its Oedipal content*, namely, the *Freudian unconscious, cannot play that role, for it is only the residue of a language distorted from the beginning*, which, once eliminated from the field of consciousness, continues to turn blindly upon itself in its teratological structure, occasionally reappearing under the phantom-like disguises of the dream or neurosis.

If we return to anthropogenesis, we can rediscover the antinomy of these two paths in the ambivalence of the social relations which developed in the endogamic community of the Mousterian period. Indeed, on the one hand, owing to the lack of women, the '*Sons*' found themselves sexual rivals of the '*Fathers*', but on the other hand, because of the communal economy which remained dominant, the immediate communal relation maintained between them an identification without jealousy, according to the inherited tradition of the original community of the Chellean. The same was undoubtedly true for the '*Mothers*' who, on the one hand, because of their age, appeared as objects of desire to the '*Sons*' but, on the other hand, as mothers responsible for the fire-place and guardians of precious provisions within the framework of the community, could not fail to elicit both respect and admiration in them. And it was undoubtedly that respectful identification of the '*Sons*' with their social parents which initiated them into the practice of developing personal relationships. In fact, it was probably with the birth of the pairing household during the Mousterian that the first *inter-subjective relations* were constituted at the heart of the immediate communal relation and on the foundation of the new division of labor between man and woman practiced in the form of the couple. While the still undifferentiated communal economy of the Chellean allowed only for an essentially collective identification relation, the division of labor in the pairing household created a new relation implying difference within identity, since each spouse had a *personal responsibility toward the other* for the economic well-being of the common household. The reciprocal affirmation of this *first personal relation of production*, as *the specific individualized form of the social relation of production*, modeled in the language of real life the double symmetrical gesture by which each speaker pointed his finger to his hearer and, correlatively, returned it upon himself, a sign which projects as tendential meaning the reciprocal image of the 'you' and the 'me'. It is this first *personal language* which, by being reflected in the inner language of consciousness, raised the subject to the ideal form of *personal existence*. And it is this *original form of personality*, constituted first of all in the '*Fathers*' and '*Mothers*' of the Neanderthal community that was transmitted to the '*Sons*' by the education of language in the relations of respectful identification which developed parallel to the distorted relations of the 'Oedipal' triangle.

This hypothesis of the birth of human personality during the Mousterian can be confirmed by the fact that it is precisely at that time that the first graves appear. It is clear that the burying of the dead presupposes an ideal representation of the human subject, posited as existing in a more or less independent manner of the state of his body. The belief in the continual existence of the soul when the body becomes lifeless and slowly decomposes, appears here as an exaggerated development of that idealized image of the *personal subject*, as it was originally constituted in language and consciousness on the foundation of the *personal relations of production* which appeared within the framework of communal relations with the origin of the pairing family.

Thus when the child at about age three, at the age of Neanderthal man, reconstitutes within himself this first form of personality according to the hereditary dialectic of prehistoric linguistic structures, he is not helped in any way, but only hindered, by the *Oedipal crisis* which emerges within him at the same time. The jealousy of the exclusive desire, which blindly repeats the semantic distortion of the Neanderthal Oedipus, and distorts it even more, comes into contradiction which the real aspirations of his budding personality, which can mature only by means of affectionate identification, realizing the reciprocity of the *you* and the *me* with *both* of his parents. The Oedipus complex is thus in no way to be found at the origin of the child's self: it is opposed from the beginning to the most profound structure of personal existence, and thus opens the way to neurosis which destroys the self. It is *alienated language*, which quickly shuts itself up in its own powerlessness, in order to stammer indefinitely in the vampire-like impasse of dreams and nightmares, of obsessions and hallucinations. Turning back from child prehistory to the prehistory of humanity enables us to clarify the pathological and essentially verbal nature of the Oedipal phantasmagoria: *language was incestuous and the Oedipal crisis linguistic*.

NOTES

For the references to Engels, Freud, Lenin, Marx and Piaget in the Notes that follow, the editions listed below are the ones cited. Complete information concerning all other works cited is given in the Notes.

ENGELS, FRIEDRICH

Dialectics of Nature. Translated and edited by Clemens Dutt. New York: International Publishers, 1940.

Dialectique de la nature. Paris: Editions sociales, 1971.

Herr Eugen Dühring's Revolution in Science (Anti-Dühring). Translated by E. Burns, edited by Clemens Dutt. New York: International Publishers, c. 1939, 1966.

M. E. Dühring bouleverse la science (anti-Dühring). Translated by E. Bottigelli. 3rd ed. rev. Paris: Editions sociales, 1972.

The Origin of the Family, Private Property and the State. Translated by A. West (and revised for this edition). New York: International Publishers, 1972.

L'origine de la famille, de la propriété privée et de l'état. Translated by J. Stern. Paris: Editions sociales, 1971.

FREUD, SIGMUND

The Standard Edition of the Complete Psychological Works of Sigmund Freud. (*S.E.*). Translated from the German under the General Editorship of James Strachey, in collaboration with Anna Freud, assisted by Alix Strachey and Alan Tyson. 24 vols. London: Hogarth Press and The Institute of Psycho-Analysis, 1953–1974:

'The Dissolution of the Oedipus Complex (1924),' *S.E.* **19** 173–179.

'New Introductory Lectures on Psycho-Analysis (1933 [1932]),' *S.E.* **22** 5–182.

'An Outline of Psycho-Analysis (1940 [1938]),' *S.E.* **23** 144–207.

'Some Psychical Consequences of the Anatomical Distinction between the Sexes (1925),' *S.E.* **19** 248–258.

La vie sexuelle. Translated by D. Berger, J. Laplanche and others. 5th rev. and corr. ed. Paris: Presses universitaires de France. (Contains French translations of the first and fourth papers listed immediately above.)

LENIN, V. I.

Collected Works. (*C. W.*) 47 vols (including 2 index vols). London: Lawrence and Wishart; Moscow: Foreign Languages Publishing House, 1960–1980. ("This English edition of Lenin's *Collected Works* is a translation of the fourth, enlarged Russian edition prepared by the Institute of Marxism-Leninism, Central Committee of the C.P.S.U. Corrections have been made to some of the texts and notes in accordance with the fifth Russian edition . . . ")

Oeuvres. Paris: Editions sociales; Moscow: Editions en langues étrangères, 1958–.

'What the 'Friends of the People" Are' (1894), *CW* **1** (1960) 129–332.

Materialism and Empirio-Criticism (1908), *CW* **14**, 1962.
Matérialisme et empiriocriticisme (1908), *Oeuvres* **14**, 1962.

Philosophical Notebooks, CW **38**, 1961.
Cahiers philosophiques, Oeuvres **38**.

Letters, February 1912–December 1922, *CW* **35**, 1966.

'The State and the Revolution' (1918), *CW* **25**, pp. 381–492.

MARX, KARL

Capital. Vol. 1, translated by B. Fowkes, vols. 2 and 3, by D. Fernbach. 3 vols. New York: Vintage Books, 1977–1981.

Das Kapital. Edited by Horst Merbach. 3 vols. Berlin: Dietz, 1966–68. This edition is identical with vols. 23–25 of Karl Marx, Friedrich Engels, *Werke*. 39 vols. (plus 2 vol. supp.), Berlin: Dietz, 1957 –.

Le capital. (various translators) 3 vols. in 8. Paris: Editions scoiales, 1950–1960.

A Contribution to the Critique of Political Economy. Translated by N. I. Stone. Chicago: Kerr, 1913.

Contribution à la critique de l'économie politique. Translated by M. Husson and G. Badia. Paris: Editions sociales, 1957.

Critique of Hegel's 'Philosophy of Right'. Edited by Joseph O'Malley. Translated by A. Jolin and J. O'Malley. Cambridge: Cambridge University Press, 1970.

Grundrisse, Foundations of the Critique of Political Economy. Translated by M. Nicolaus. New York: Vintage Books, 1973.

Grundrisse der Kritik der politischen Ökonomie (*Rohentwurf*) (1939). Berlin: Dietz, 1953.

Fondements de la critique de l'économie politique. Translated by R. Dangeville. 2 vols. Paris: Editions Anthropos, 1969.

MARX, KARL AND ENGELS, FRIEDRICH

Collected Works. (*C.W.*) New York: International Publishers; London: Lawrence and Wishart; Moscow: Progress Publishers, 1975 –:

'Economic and Philosophic Manuscripts of 1844.' *C.W.* 3 (1975) 229–346. ['Ökonomisch-philosophisch Manuscripte.' In *Kleine ökonomische Schriften*, pp. 42–166. Berlin: Dietz, 1955; *Manuscrits de 1844, économie politique et philosophie*. Translated by E. Bottigelli. Paris: Editions sociales, 1962.]

'The German Ideology.' *C.W.* 5 (1976) 19–539. [*L'idéologie allemande*. Translated R. Cartelle and G. Badia. Paris: Editions sociales, 1965.]

'The Holy Family.' *C.W.* 4 (1975) 5–211. [*La sainte famille*. Translated by E. Cogniot. Paris: Editions sociales, 1969.]

PIAGET, JEAN

The Child's Construction of Reality. Translated by M. Cook. London: Routledge and Kegan Paul, 1976. (Also: *The Construction of Reality in the Child*. Translated by M. Cook. New York: Basic Books, 1954.)

La construction du réel chez l'enfant. Neuchâtel: Delachaux and Niestlé, 1937.

The Origins of Intelligence in Children. Translated by M. Cook. New York: International Universities Press, c. 1952, 1977.

La naissance de l'intelligence chez l'enfant. 2nd ed. Neuchâtel: Delachaux and Niestlé, 1948.

Play, Dreams and Imitation in Childhood. Translated by C. Gattegno and F. M. Hodgson. New York: Norton, c. 1951, 1962.

Formation du symbole chez l'enfant. Neuchâtel: Delachaux and Niestlé, 1945.

NOTES TO THE FIRST INVESTIGATION

[1] Karl Marx, *Capital* 1, chap. 7, sect. 1, p. 284; *Das Kapital* 1, p. 193; *Le capital*, part I, vol. 1, p. 181.

[2] V. P. Iakimov, ed., *U istokov chelovechestva* [The Origins of Man] (Moscow, 1964).

[3] Friedrich Engels, *Dialectics of Nature*, p. 284; *Dialectique de la nature*, p. 175.

[4] Edward George Boulenger, *Apes and Monkeys* (New York: McBride, 1937), p. 46.

[5] V. I. Lenin, *Materialism and Empirio-criticism*, p. 69; *Matérialisme et empiriocriticisme*, p. 69.

[6] Iakimov, *op. cit.*

[7] *Ibid.*

[8] Karl Marx, *Capital* 1, chap. 1, sect. 3, p. 144n; *Das Kapital* 1, p. 67, n. 18; *Le capital*, part I, vol. 1, p. 67; n. 2. [Unfortunately this very sentence of Marx was omitted in the French edition.]

[9] Wolfgang Köhler, *The Mentality of Apes*, trans. from the 2nd rev. ed. by E. Winter (London: K. Paul, Trench, Trubner; New York: Harcourt, Brace, 1925), pp. 329–330.

[10] N. A. Tikh cited in A. Spirkin, *Proiskhozhdenie soznaniia* [The Origin of Consciousness] (Moscow, 1960), p. 59n. The original works (in Russian) of N. A. Tikh were published in his doctoral thesis, Pavlov Institute of the Academy of Sciences of the USSR, Moscow, 1950. 36 pages.

[11] Köhler, *op. cit.*, pp. 298–299.

[12] Karl Marx, *Capital* 1, chap. 1, sect. 3, p. 144n; *Das Kapital* 1, p. 67, n. 18; *Le capital*, part I, vol. 1, p. 67, n. 2.

[13] It goes without saying that this description is only fully valid under the original conditions of hominization, or the gestation period of genus *Homo*. In the second part of anthropogenesis, namely 'sapientiation' or the formation of *Homo sapiens*, the motion is internalized, to be followed by indefinitely complicated forms.

[14] Köhler, *op. cit.*, pp. 293–294.

[15] Karl Marx and Friedrich Engels, *The German Ideology*, p. 44; *L'idéologie allemande*, p. 64.

[16] *Ibid.*, p. 36; French ed., p. 50.

[17] Karl Marx and Friedrich Engels, *The Holy Family; La sainte famille.*

[18] Karl Marx and Friedrich Engels, *The German Ideology*, p. 44; *L'idéologie allemande*, p. 59.

[19] *Ibid.*, p. 36; French ed., p. 50.

[20] Karl Marx, *Capital* 1 chap. 1, sect. 3, p. 143; *Le capital*, part I, vol. 1, p. 66 (our emphasis).

[21] *Ibid.*, chap. 1, sect. 4, p. 173n; French ed., part I, vol. 1, p. 92n.

[22] *Ibid.*, chap. 1, sect. 3, p. 142n; *Das Kapital*, p. 65, n. 17a (our emphasis).

[23] Karl Marx, *Contribution to the Critique of Political Economy*, pp. 63–64; *Contribution à la critique de l'économie politique*, p. 33 (our emphasis).

[24] Spirkin, *op. cit.*

[25] V. I. Lenin, *Materialism and Empirio-criticism*, p. 51; *Matérialisme et empiriocriticisme*, p. 50.

[26] Karl Marx, *Grundrisse*, 'Notebook' V, p. 490; *Grundrisse*, Heft V, p. 390; *Fondements* 1, p. 453.

[27] Immanuel Kant, *Critique of Pure Reason*, trans. N. K. Smith (New York: St. Martin's Press, 1965), p. 167; *Kritik der reinen Vernunft* (Leipzig: Reclam, n.d.), p. 209.

[28] V. I. Lenin, *Materialism and Empirio-criticism*, p. 147; *Matérialisme et empiriocriticisme*, p. 152 (our emphasis).

[29] *Ibid.*, p. 90; French ed., p. 91 (our emphasis).

[30] *Ibid.*, p. 244; French ed., p. 253.

[31] *Ibid.*

[32] Karl Marx, *Grundrisse*, 'Notebook' V, p. 490; *Grundrisse*, Heft V, p. 390; *Fondements* 1, pp. 452–453.

[33] Karl Marx, *The German Ideology*, p. 44; *L'idéologie allemande*, p. 59.

[34] Friedrich Engels, *Dialectics of Nature*, p. 35; *Dialectique de la nature*, p. 75.

[35] *Ibid.*, p. 21; French ed., p. 43.

[36] Friedrich Engels, *Anti-Dühring*, p. 68; *M. E. Dühring bouleverse la science*, p. 393.

[37] V. I. Lenin, *Philosophical Notebooks*, p. 182; *Cahiers philosophiques*, p. 172.

[38] Friedrich Engels, *Dialectics of Nature*, p. 228; *Dialectique de la nature*, 383. Cf. *M. E. Dühring bouleverse la science*, pp. 198 and 210.

[39] Friedrich Engels, *Dialectics of Nature*, p. 17; *Dialectique de la nature*, p. 41.

[40] Karl Marx, *Contribution to the Critique of Political Economy*, p. 268; *Contribution à la critique de l'économie politique*, p. 150.

NOTES TO THE SECOND INVESTIGATION

[1] Roman Jakobson, Chapter 2 in *Results of the Conference of Anthropologists and Linguists*, by Claude Lévi-Strauss, Roman Jakobson, C. F. Voegelin, and Thomas A. Sebeok, Indiana University Publications in Anthropology and Linguistics Memoir 8 and International Journal of American Linguistics Memoir 8 (April 1953), p. 21. The French version is 'Le langage commun des linguistes et des anthropologues,' in Roman Jakobson, *Essais de linguistique générale*, trans. N. Ruwet (Paris: Les Editions de Minuit, 1963), p. 41.

[2] *Ibid.*; French ed., p. 42.

[3] *Ibid.*; French ed., pp. 41–42.

[4] V. I. Lenin, cited from the French translation of 'What the "Friends of the People" Are.'

[5] V. I. Lenin, *Philosophical Notebooks*, p. 274 (our emphasis).

[6] V. I. Lenin, *Materialism and Empirio-criticism*, pp. 260–261; *Matérialisme et empiriocriticisme*, p. 271.

[7] *Ibid.*, p. 130; French ed., p. 133 (our emphasis). (Note that the French translation of the fourth edition has *témoignage* [lit.: testimony, evidence] instead of *les indications* [information].)

[8] *Ibid.*, p. 267; French ed., p. 278.

[9] Karl Marx, *Capital* 1, postface to the 2nd ed., p. 102; *Le capital*, part I, vol. 1, p. 29.

[10] *Ibid.*

[11] See: Trân Duc Thao, 'Le mouvement de l'indication comme forme originaire de la conscience,' *La Pensée*, no. 128, August 1966. The traditional designation of Australopithecus becomes difficult to maintain from the moment that this group is separated from the Pongid family in order to connect it to that of the Hominids. Bounak suggests Protanthropus and Leroi-Gourhan, Australanthropus. In what follows we will generally use the name of Australanthropi since they seem to have developed primarily in the southern and western part of Africa.

[12] It is essential to note that the fundamental change does not consist here in the general acquisition of bipedal gait, but of *bipedal gait insofar as it liberates the hand*. Thus the gibbons walk on two feet but they are obliged to extend their arms to keep their balance: thus there is no liberation of the hand.

The *qualitative leap* which was realized in the transition from the anthropoid to the Australanthropus, resulted from a long *quantitative* development deriving from the manipulation and use of the instrument which, by occupying the ape's hands, obliged him more and more to stand on his feet. The adaptation to erect gait resolved the growing contradiction between the beginning of labor and the slouching gait in the anthropoid ancestor: it is such a dialectic which makes for the distinction between the bipedal gait of Australanthropus and that of the kangaroo, for example, or of the penguin etc. whose fore-limbs are, so to speak, '*free*' but not 'liberated'. "For erect gait among our hairy ancestors to have become first the rule and in time a necessity," says Engels, "presupposes that in the meantime the hands became more and more devoted to other functions . . . Many monkeys use their hands to build nests for themselves in the trees or even, like the chimpanzee, to construct roofs between the branches for protection against the weather. With their hands they seize hold of clubs to defend themselves against enemies, or bombard the latter with fruits and stones" (Friedrich Engels, *Dialectics of Nature*, p. 280; *Dialectique de la nature*, p. 172).

13 According to Leakey, the fauna of Olduvai Gorge presents "affinities with the upper Villefranchien rather than the lower" (Louis S. B. Leakey, *Olduvai Gorge, 1951–61*, by L. S. B. Leakey with contributions by P. M. Butler and others, 3 vols. (Cambridge: Cambridge University Press, 1965 –), vol. 1, *A Preliminary Report on the Geology and Fauna* (1965), p. 74).

Tobias also places *Homo habilis*, as well as the Australanthropi whose bones we possess, in the second part of the lower Pleistocene. Consequently, the Australanthropus ancestor, from whom *Homo habilis* originated, had to develop in the first part of the lower Pleistocene, beginning perhaps toward the end of the Pliocene (Tobias, 'Early Man in East Africa,' *Science* **149**, no. 3679, 1965). These determinations seem acceptable to us independently of any considerations about absolute dates. The latter still remain rather uncertain, but are not indispensable for a general representation of the whole development.

The majority of present taxonomies agree to separate the Australanthropi from the *Pongid* family in order to link them to that of the Hominids (Bounak, Heberer, Le Gros, Clark, Woo Ju-Kang, Genet-Varcin, Lerori-Gourhan, Königswald). The major reason for this is that the liberation of the hand resulting from the adaptation of the foot to erect posture, constitutes the *decisive step* which opens the way to *hominization*. It was precisely this idea which Engels presented at the end of the last century: " . . . these apes when moving on level ground began . . . to adopt a more and more erect posture in walking. This *was the decisive step in the transition from ape to man – the hand became free*" (Friedrich Engels, *Dialectics of Nature*, pp. 279–281; *Dialectique de la nature*, pp. 172–173).

Since there could be no question of classifying the Australopitheci, or Australanthropi in the genus *Homo*, since they only knew the work of adaptation, and not of production, we propose to consider them as defining a *Praehomo* genus, which appeared toward the boundary between the Tertiary and Quaternary, whose presently known specimens are but late representatives. Prehominid development, strictly speaking, which prepared for the appearance of genus *Homo* in the form of *Homo habilis*, must be placed in the *first part of the lower Pleistocene* with perhaps a notch in the end of the Pliocene: this is the *intermediary stage* which leads to the transition from ape to man. In fact, the Australanthropi do indeed correspond to what Engels called "these transitional beings" who, having transcended animality, strictly speaking, by the decisive progress which was the freeing of the hand, could gradually raise the level of adaptive work, without yet reaching the form of production characteristic of human existence: "At first, therefore, the operations, for which our ancestors gradually learned to adapt their hands during the many thousands of years of transition from ape to man, could only have been very simple. The lowest savages, even those in whom a regression to a more animal-like condition, with a simultaneous physical degeneration, can be assumed to have occurred, are nevertheless far superior to these transitional beings. Being the first flint could be fashioned into a knife by human hands, a period of time must probably have elapsed in comparison with which the historical period known to us appears insignificant. But the decisive step was taken: *the hand became free* and could henceforth attain ever greater dexterity and skill, and the greater flexibility thus acquired was inherited and increased from generation to generation" (Friedrich Engels, *Dialectics of Nature*, p. 281; *Dialectique de la nature*, pp. 172–173).

[14] Karl Marx, *Capital* **1**, chap. 7, sect., 1, p. 284; *Le capital*, part I, vol. 1, p. 181.

[15] Mrs. M. D. Leakey, 'A Review of the Oldowan Culture from Olduvai Gorge, Tanzania,' *Nature* **210**, no. 5035 (1966), p. 463.

[16] Georg Wilhelm Friedrich Hegel, *Phenomenology of Spirit*, trans. A. V. Miller (Oxford: Oxford University Press, c. 1977, 1981), p. 12; *Phénoménologie de l'esprit*, trans. J. Hyppolite, 2 vols. (Paris: Aubier, Editions Montaigne, 1939–41), vol. 1, p. 20.

[17] Louis S. B. Leakey, P. V. Tobias, J. R. Napier, 'A New Species of the Genus *Homo* from Olduvai Gorge,' *Nature* **202**, no. 4927 (1964), pp. 7–9. See the discussion in *Voprosy antropologii* [Problems in Anthropology], fasc. 19 (Moscow, 1965); *Fossil Hominids and the Origin of Man* (in Russian) (Moscow, 1966); Iurii Georgievich Reshetov, *Priroda Zemli i proiskhozhdenie cheloveka* [The Nature of the Earth and the Origin of Man] (Moscow, 1966).

[18] *Voprosy antropologii* [Problems in Anthropology], fasc. 19 (Moscow, 1965), pp. 9–10 and 24.

[19] Karl Marx, *Capital* **1**, chap. 7, sect. 1, p. 284; *Le capital*, part I, vol. 1, p. 181.

[20] Henri Wallon, *De l'acte à la pensée, essai de psychologie comparée* (Paris: Flammarion, 1942), p. 78.

[21] Trân Duc Thao, *op. cit.*

[22] V. P. Iakimov, 'The Australopitheci,' in *Fossil Hominids and the Origin of Man* (in Russian) (Moscow, 1966), pp. 74–76. The famous long bones whose pointed fragments could have served as daggers were probably broken on the ground or on a rock, which does not go beyond the limit of an act of direct manipulation. In the case where they would have been broken by means of a stone, it would only have been for the purpose of extracting the bone marrow; thus it would only be a question of an entirely ordinary act of the use of the instrument for the satisfaction of biological need. Once this need was satisfied, the remaining bone fragments then functioned only as natural objects. Thus, if afterwards they function as daggers to attack game, it will only be as natural instruments. It is true that they can still be prepared when needed, but we have no reason to assume that this was done through the intermediary of another instrument. Thus in the case of the two bone fragments found together, one fitting in the medullar cavity of the other, it is evident that we are faced, if not with a simple chance event, then with the result of an operation which in no way goes beyond the fitting of two reeds together as in the case of the chimpanzee.

[23] V. P. Iakimov, in *Voprosy antropologii* [Problems in Anthropology], fasc. 19 (Moscow, 1965), p. 9.

[24] *The Peoples of Australia and Oceania* (in Russian) (Moscow), p. 275.

[25] Karl Marx, *Capital* **1**, chap. 7, sect. 1, p. 285; *Le capital*, part I, vol. 1, p. 181: "An instrument of labour is a thing, or a complex of things, which the worker interposes between himself and the object of his labour and which serves as a conductor, directing his activity onto that object."

[26] Nadezhda Nikolaevna Ladygina-Kots, 'Instrumental Activity of Apes and the Problem of Anthropogenesis,' in *Contemporary Anthropology* (in Russian) (Moscow, 1964), p. 141.

27 Henriette Alimen, *The Prehistory of Africa*, trans. A. H. Brodrick (London: Hutchinson, 1957), p. 205; *Préhistoire de l'Afrique* (Paris: N. Boubée, c. 1955, 1966), p. 245. Trân Duc Thao used the Russian translation (Moscow, 1960), p. 236.

28 *Ibid.*, p. 282; French ed., p. 334; Russian ed., p. 314.

29 Karl Marx, *Capital* 1, chap. 7, sect. 1, p. 284; *Le capital*, part I, vol. 1, p. 181.

30 Friedrich Engels, *Dialectics of Nature*, p. 283; *Dialectique de la nature*, p. 177.

31 *Ibid.*, p. 285; French ed., p. 175.

32 Karl Marx, *Capital* 1, chap. 7, sect. 1, p. 286; *Le capital*, part I, vol. 1, pp. 182–183.

33 Friedrich Engels, *Dialectics of Nature*, p. 291; *Dialectique de la nature*, p. 180.

34 It is obviously a question here only of the general level of behavior, characterized by the capacity for solving practical problems by trial and error. For concrete performances, it is evident that the anthropoid has the better of the child owing to his agility and his sensori-motor experience. In particular, it is through the experience of previous attempts that we must explain the apparent phenomena of the reorganization of the perceptive field in the case of the chimpanzee's behavior in handling instruments, which Köhler has described as being purely intuitive (see the refutation of Köhler's theses in Ladygina-Kots, *op. cit.*).

35 Jean Piaget, *The Child's Construction of Reality*, pp. 79–82; *La construction du réel chez l'enfant*, pp. 70–72. Thérèse Gouin-Décarie, *Intelligence and Affectivity in Early Childhood: an Experimental Study of Jean Piaget's Object Concept and Object Relations*, trans. E. P. Brandt and L. W. Brandt (New York: International Universities Press, 1965), pp. 155–158; *Intelligence et affectivité chez le jeune enfant; étude expérimentale de la notion d'object chez Jean Piaget et de la relation objectale* (Neuchâtel: Delachaux and Niestlé, 1962), pp. 148–150.

36 Jean Piaget, *Play, Dreams and Imitation in Childhood*, p. 217; *La formation du symbole chez l'enfant*, p. 231.

37 Aleksandr Nikolaevich Gvozdev, *Voprosy izucheniia detskoi rechi* [Problems in the Study of Child Language] (Moscow, 1961), p. 162.

38 Édouard Pichon, *Le développement psychique de l'enfant et de l'adolescent* (Paris: Masson, 1936), p. 59.

39 Jean Piaget, *Play, Dreams and Imitation in Childhood*, p. 218; *La formation du symbole chez l'enfant*, p. 232.

40 G. L. Rosengart-Pouklo, *Formirovanie rechi o deteij rannero vozrasta* [Language Formation in Young Children] (Moscow, 1963, cited in Solomon Davidovich Katsnel'son, *Soderzhanie slova* [The Content of the Word] (Moscow, 1965), p. 29.

41 Jean Bourjade, *Études de psychologie de l'enfant* (Paris: Les Belles Lettres, 1962), p. 33.

42 Pichon, *loc. cit.*

43 Aleksandr Georgievich Spirkin, 'Iazyk – sredstvo obmena myshliami,' in *Myshlenie i iazyk* [Language – Its Role in the Formation of Thought in *Thought and Language*] (Moscow, 1958), pp. 36–37.

44 V. I. Kotchetkova, 'The Evolution of Specifically Human Areas in the Hominid Cerebral Cortex' (in Russian), *Voprosy antropologii* [Problems in Anthropology], fasc. 7 (Moscow, 1961), p. 16.

45 V. I. Kotchetkova, 'Comparative Characteristics of Hominid Endocrania from the Paleoneurological Point of View' (in Russian), in *Fossil Hominids and the Origin of Man* (in Russian) (Moscow, 1966), p. 490.

[46] Tatiana Efimovna Konnikova, *Nachalnyii etap v razvitii rechi* [The Initial Stage in Language Development] (Leningrad, 1947). Summarized by Daniil Borisovich El'konin, *Detskaia psikhologiia* [Child Psychology] (Moscow, 1960), p. 99.

[47] We shall not repeat here the details which have already been presented about the ape's original cognizance of the indicative sign.

[48] Jean Piaget, Observations 58–59 in *The Child's Construction of Reality*, pp. 71–72; in *La construction du rèel chez l'enfant*, pp. 64–65.

[49] *Ibid.*, Observations 60–63, pp. 74–77; in French ed., pp. 66–69.

[50] *Ibid.*, Observations 64–66, pp. 79–82; in French ed., pp. 70–72.

[51] Gouin-Décarie, *op. cit.*, pp. 151–153; French ed., 143–145.

[52] Karl Marx, *Capital* 1, chap. 13, p. 445; *Le capital*, part I, vol. 2, p. 20.

[53] *Ibid.*, p. 464. [Unfortunately this English translation does not mention Briareus, a hundred-handed monster, son of Heaven and Earth. See *Iliad* I, 396 ff. – Ed.] *Le capital*, part I, vol. 2, p. 35: "Briarée, dont les mille mains sont armées d'outils divers . . ."

[54] Jean Piaget, Observation 101(a), in *Play, Dreams and Imitation in Childhood*, p. 216; in *La formation du symbole chez l'enfant*, pp. 230–231.

[55] *Ibid.*, Observation 102, pp. 217–218; in French ed., pp. 231–232.

[56] Arnold Gesell and Frances L. Ilg, *Infant and Child in the Culture of Today* (New York: Harper, 1949), p. 133. Gesell places the beginning of drawing at 15 months, but this is undoubtedly a minimal limit.

[57] Jean Piaget, Observation 52, in *Play, Dreams and Imitation in Childhood*, p. 63; *La formation du symbole chez l'enfant*, p. 64.

[58] Pichon, *op. cit.*, p. 60.

[59] Odette Brunet and Irène Lézine, *Le développement psychologique de la première enfance* (Paris: Presses universitaires de France, 1951), p. 64.

[60] Jean Piaget, Observation 101(a), in *Play, Dreams and Imitation in Childhood*, pp. 216–217; *La formation du symbole chez l'enfant*, pp. 230–231.

[61] In little Génia studied by Gvozdev, the stage of the functional sentence goes from 19 to 21 months.

[62] The question can be raised as to how an object which is practically always at rest, like a trunk, can be supposed to be in motion. The answer is that the indicative gesture which aims at the trunk can only be developed according to a scheme of action previously established concerning more or less analogous objects, but in motion.

[63] Jean Piaget, Observation 104, in *Play, Dreams and Imitation in Childhood*, p. 222; *La formation du symbole chez l'enfant*, pp. 236–237.

[64] Jean Piaget, *The Child's Construction of Reality*, p. 346; *La construction du rèel chez l'enfant*, p. 304.

[65] Ladygina-Kotz, *op. cit.*, pp. 145–147. Köhler also makes this point in his account, but he did not draw out all of its theoretical consequences.

[66] Jean Piaget, *The Origins of Intelligence in Children*, pp. 337–338; *La naissance de l'intelligence chez l'enfant*, pp. 293–294. We saw earlier in the observations on the beginnings of representation that Lucienne exhibited behaviors between 13 and 15 months which Jacqueline did not achieve until 18 or 19 months. Since Jacqueline's whole development seems to have been nearly normal, we may suppose that Lucienne's 16 months of age corresponds more or less to the level of 19 months.

[67] Piaget's text says that the child "moved the forefinger of her right hand an inch or

so away from her thumb." Since, however, she probably holds her hand in that position for an instant, just the time it takes to repeat: "*Little, little* – " we can just as well consider the signifying gesture in the form of having the thumb nearer to the index finger, a drawing nearer which is projected on the signified image in the form of making smaller the represented object, as it appears in its (supposed) motion. (Cf. Jean Piaget, Observation 104, *Play, Dreams and Imitation in Childhood*, p. 222; *La formation du symbole chez l'enfant*, pp. 236–237.)

68 Jean Piaget, *The Child's Construction of Reality*, p. 296; *La construction du réel chez l'enfant*, pp. 258–259.

69 Jean Piaget, *Play, Dreams and Imitation in Childhood*, p. 226; *La formation du symbole chez l'enfant*, pp. 239–240.

70 Cf. Daniil Borisovich El'konin, *Detskaia psikhologiia* [Child Psychology] (Moscow, 1960), p. 101.

71 Louis S. B. Leakey, P. V. Tobias, J. R. Napier, 'A New Species of the Genus *Homo* from Olduvai Gorge,' *Nature* **202**, no. 4927 (1964), p. 9.

72 The question can be raised as to whether the construction of these embankments can already be considered as a work of elaboration since the Australanthropi probably manipulated stones with their hands, without any instrument, which apparently sends us back to a simple direct act of the manipulation of raw material. Actually, it is these stones themselves, which, starting with the second layer above the ground, alternately function as instrument and as material. In effect, the stones of the first layer put on the ground must be fixed in that position precisely by the weight of those of the second layer, which consequently functions as instrument at the moment that the subject puts them on the first ones. They, in their turn, function as material in relation to the third layer and so on. It is in such a work that the form of alignment is realized as the instrumental form of the embankment.

73 Karl Marx, *Capital* 1, chap. 7, sect. 1, p. 284; *Le capital*, part I, vol. 1, p. 181.

74 *Ibid*., p. 287; *Das Kapital* 1, p. 195.

75 *Ibid*., postface to the 2nd ed., p. 103; *Le capital*, part I, vol. 1, p. 29.

76 V. I. Lenin, *Philosophical Notebooks*, p. 111 (marginal note).

77 Friedrich Engels, *Dialectics of Nature*, p. 328; *Dialectique de la nature*, p. 234.

78 V. I. Lenin, *Materialism and Empirio-criticism*, p. 270.

79 Georg Wilhelm Friedrich Hegel, *Phenomenology of Spirit*, trans. A. V. Miller (Oxford University Press, c. 1977, 1981), p. 27; *Phénoménologie de l'esprit*, trans. J. Hyppolite, 2 vols. (Paris: Aubier, Editions Montaigne, 1939–41), vol. 1, p. 40.

80 V. I. Lenin, *Philosophical Notebooks*, pp. 361–362.

81 Karl Marx, *The Economic and Philosophic Manuscripts of 1844*, p. 276; *Ökonomisch-philosophische Manuscripte*, p. 104; *Manuscrits de 1844, économie politique et philosophie*, p. 63.

82 Karl Marx, *Capital* 1, chap. 7, sect. 1, p. 283; *Le capital*, part I, vol. 1, p. 180.

83 *Ibid*., pp. 283–284; French ed., p. 178.

84 Bounak, 'Language and Intelligence: The Stages of Their Development in Anthropogenesis,' *Fossil Hominids and the Origin of Man* (in Russian) (Moscow, 1966), p. 538.

NOTES TO THE THIRD INVESTIGATION

[1] André Green. [The source has not been identified. Green is the author of *Le discours vivant: la conception psychanalytique de l'affect* (Paris: Presses universitaires de France, 1973), and *Un OEil en trop, le complexe d'OEdipe dans la tragédie* (Paris: Editions de minuit, 1969) – Ed.]

[2] Thérèse Gouin-Décarie, *Intelligence and Affectivity in Early Childhood; An Experimental Study of Jean Piaget's Object Concept and Object Relations*, trans. E. P. Brandt and L. W. Brandt (New York: International Universities Press, 1965), pp. 116–117; *Intelligence et affectivité chez le jeune enfant; étude expérimentale de la notion d'object chez Jean Piaget et de la relation objectale* (Neuchâtel: Delachaux and Niestlé, 1962), p. 111.

[3] Sigmund Freud, 'Some Psychical Consequences of the Anatomical Distinction between the Sexes,' p. 251; *La vie sexuelle*, p. 126.

[4] *Ibid.*, p. 256; French ed., p. 130 (our emphasis).

[5] *Ibid.*, p. 250; French ed., p. 125.

[6] Sigmund Freud, 'An Outline of Psychoanalysis,' p. 150–151; cf. *La vie sexuelle*, p. 83.

[7] Friedrich Engels, *The Origin of the Family, Private Property and the State*, pp. 98–99; *L'origine de la famille, de la propriété privée et de l'état*, pp. 42–43.

[8] *Ibid.*, pp. 99–100; French ed., p. 43.

[9] *Ibid.*, pp. 100–101; French ed., p. 44.

[10] Friedrich Engels, *Dialectics of Nature*, p. 283; *Dialectique de la Nature*, p. 174.

[11] V. I. Lenin, *CW* **35**, pp. 128–129.

[12] V. I. Lenin, 'The State and the Revolution,' *CW* **35** pp. 389–390.

[13] See the bibliography and analysis of opinions in Semenov, *How Mankind was Born* (in Russian) (Moscow), pp. 18–33.

[14] Franck Bourdier, *Préhistoire de France* (Paris: Flammarion, 1967), pp. 193, 195, 215.

[15] Friedrich Engels, *The Origin of the Family* . . . , p. 137; *L'origine de la famille* . . . , p. 63.

[16] Baldwin Spencer and Frances James Gillen, *The Native Tribes of Central Australia* (New York: Dover, 1968, c. 1899), pp. 60 ff.

[17] For simplicity of exposition, we are leaving out the rule according to which *nupa* can be only female cousins from the elder brothers of the mother's side or of the elder sisters of the father's side.

[18] Cited by Semenov, *op. cit.*, p. 130.

[19] Henri V. Vallois, 'The Social Life of Early Man: the Evidence of Skeletons,' in *Social Life of Early Man*, ed. Sherwood Larned Washburn (Chicago: Aldine, 1961), p. 225, table 4.

[20] On the other hand, it seems very difficult to use the collections of the lower Paleolithic. The composition of the collection of the Sinanthropi, ten men for three

women, cannot correspond to the real demographic situation for one can hardly see how, under these conditions, the group could normally reproduce. As for the collection of Ngang dong, four men and six women, according to general opinion, it represents the remnants of a ritual cannibalism. Königswald considers that this deposit constitutes a sort of "alter of skulls" comparable to those that can still be seen today in Melanesia.

Be that as it may, it is certain, as we shall show later, that the demographic disequilibrium between the sexes observed during the Mousterian, already existed during the lower Paleolithic.

[21] Vallois, *op. cit*., p. 224, table 3.

[22] Cf. P. I. Borisovsky, 'The Problems of the Formation of Human Society and the Archaeological Discoveries of the Last Ten Years,' in *Lenin's Ideas in the Study of the History of Primitive Society, Slave-Owning Society and Feudalism* (in Russian) (Moscow), p. 72.

[23] Friedrich Engels, *The Origin of the Family* . . . , p. 100; *L'origine de la famille* . . . , p. 40.

[24] Spencer and Gillen, *op. cit*., pp. 558–559.

[25] Cf. Sherwood Larned Washburn, 'Tools and Human Evolution,' and John Napier, 'The Antiquity of Human Walking,' in *Human Variation and Origins; an Introduction to Human Biology and Evolution (Readings from Scientific American)* (San Francisco: W. H. Freeman, 1967).

[26] Napier, *op. cit*., p. 126.

[27] Vallois, *op. cit*., pp. 223, 228.

[28] M. D. Leakey, 'A Review of the Oldowan Culture from Olduvai Gorge, Tanzania,' *Nature* **210**, no. 5035 (1966), pp. 462–463. The reproduction of the proto-biface is in vol. 212, no. 5062, p. 579 (in Mrs. Leakey's article, 'Primitive Artefacts from Kanapoi Valley').

[29] Friedrich Engels, *The Origin of the Family* . . . , p. 255.

[30] For this whole section, cf. Semenov, *op. cit*., pp. 191–195.

[31] Alberto C. Blanc, 'Some Evidence for the Ideologies of Early Man,' in *Social Life of Early Man*, ed. Sherwood Larned Washburn (Chicago: Aldine, 1961), p. 133.

[32] Sigmund Freud, 'The Dissolution of the Oedipus Complex,' p. 184; *La vie sexuelle*, p. 118.

[33] Karl Abraham, 'Manifestations of the Female Castration Complex,' in *Selected Papers of Karl Abraham*, trans. by D.Bryan and A. Strachey, The International Psycho-analytical Library, 13 (London: Hogarth Press, 1973, c. 1927), p. 341; 'Äußerungsformen des weiblichen Kastrationkomplexes,' in *Psychoanalytische Studien zur Charakterbildung und andere Schriften*, ed. Johannes Cremerius, 2 vols. (Frankfurt am Main: S. Fischer, 1969–1971), vol. 2: *Psychoanalytische Studien* (1971), p. 72; *Oeuvres complètes*, trans. I. Barande, 2 vols. (Paris: Payot, 1965–66), vol. 2: *1913–1925. Développement de la libido, formation du caractère, études cliniques* (1966), pp. 103–104.

[34] Bourdier, *op. cit*., p. 174.

[35] Abraham, *op. cit*., p. 355; German ed., p. 85; French ed., p. 115.

[36] From the point of view of anthropogenesis, the sexual abstinence imposed upon the young Neanderthals could have favored 'youthfulness' where several authors see one of the causes for the transition from certain progressive Neanderthal forms to *Homo sapiens*. "If the traits of the infantile form had persisted into maturity, then through the morphology of its mandible and the design of its encephalon, the Neanderthal

child of La Chaise would probably have become somewhat similar to modern man." (J. Piveteau, quoted in F. Bourdier, *Préhistoire de France* (Paris: Flammarion, 1967), p. 206). It is of course understood that like fetation, youthfulness could play only a role of acceleration, the essential cause of the sapientiation, or formation of *Homo sapiens* evidently remaining the development of labor and language.

37 Vallois, *op. cit.*, p. 230.

38 *Ibid.*, p. 224.

39 Jean Laplanche and J.-B. Pontalis, *The Language of Psycho-analysis*, trans. D. Nicholson-Smith, The International Psycho-analytical Library, 94, (London: Hogarth Press, 1973), p. 57; *Vocabulaire de la psychanalyse* (Paris: Presses universitaires de France, 1967), p. 76.

40 Carl Strehlow, *Die Aranda- und Loritja-stämme in Zentral Australien*, 5 vols. in 7 (Frankfurt-a-M.: J. Baer, 1907–20); vol. 4, part 1, *Das soziale Leben der Aranda- und Loritja-stämme*, pp. 10–12.

41 Spencer and Gillen, *op. cit.*, p. 246.

42 *Ibid.*, pp. 398–399.

43 Sigmund Freud, 'The Dissolution of the Oedipus Complex,' p. 175; *La vie sexuelle*, p. 119.

44 Sigmund Freud, 'Some Psychical Consequences of the Anatomical Distinction between the Sexes,' p. 252; *La vie sexuelle*, p. 127.

45 Sigmund Freud, 'The Dissolution of the Oedipus Complex,' p. 174; *La vie sexuelle*, p. 117.

46 *Ibid.*, p. 178; French ed., p. 121.

47 Sigmund Freud, 'Some Psychical Consequences of the Anatomical Distinction between the Sexes,' p. 256–257; *La vie sexuelle*, p. 130.

48 Sigmund Freud, 'The Dissolution of the Oedipus Complex,' p. 178. *La vie sexuelle*, p. 122.

49 First quotation, Sigmund Freud, 'Some Psychical Consequences of the Anatomical Distinction between the Sexes,' p. 257; *La vie sexuelle*, p. 131; second quotation, Freud, 'The Dissolution of the Oedipus Complex,' p. 179; *La vie sexuelle*, p. 122.

50 ['Aurignacian' (from the cave of Aurignac, Haute-Garonne, France) refers to the earliest phase of European prehistoric art, i.e., the art of the Upper Paleolithic, dating from 30 000 B.C. 'Venus' is the art historical term used to designate the small, female statuary of this period. Since, with some of these figurines, breasts and abdomen are given great prominence (in some cases the sculpture actually begins at the waist), it is thought that the 'Venuses', as symbols of female fecundity, served as magical representations. (Cf. s. v. 'Prehistory,' *Encyclopedia of World Art* (New York: McGraw-Hill, 1966), *passim*.)]

51 Spencer and Gillen, *op. cit.*, p. 388–420.

52 Vladimir Rafailovich Kabo, *Proiskhozhdenie i ranniaia istoriia aborigenov Avstralii* [The Origin and Early History of the Aborigines of Australia] (Moscow, 1969).

53 Sigmund Freud, 'An Outline of Psychoanalysis,' p. 148.

54 Sigmund Freud, 'New Introductory Lectures on Psychoanalysis,' p. 95; cited in LaPlanche and Pontalis, *op. cit.*, p. 216; French ed., p. 362.

55 Sigmund Freud, 'Some Psychical Consequences of the Anatomical Distinction between the Sexes,' p. 250.

INDEX OF NAMES

GPSR Compliance

The European Union's (EU) General Product Safety Regulation (GPSR) is a set of rules that requires consumer products to be safe and our obligations to ensure this.

If you have any concerns about our products, you can contact us on ProductSafety@springernature.com

In case Publisher is established outside the EU, the EU authorized representative is:

Springer Nature Customer Service Center GmbH
Europaplatz 3
69115 Heidelberg, Germany

Batch number: 10823811

Printed by Printforce, the Netherlands